EXPANDING BIOFUEL PRODUCTION AND THE TRANSITION TO ADVANCED BIOFUELS

LESSONS FOR SUSTAINABILITY FROM THE UPPER MIDWEST

SUMMARY OF A WORKSHOP

Patricia Koshel and **Kathleen McAllister**
Rapporteurs

Science and Technology for Sustainability Program
Policy and Global Affairs

NATIONAL RESEARCH COUNCIL
OF THE NATIONAL ACADEMIES

THE NATIONAL ACADEMIES PRESS
Washington, D.C.
www.nap.edu

THE NATIONAL ACADEMIES PRESS 500 Fifth Street, N.W. Washington, D.C. 20001

NOTICE: The project that is the subject of this report was approved by the Governing Board of the National Research Council, whose members are drawn from the councils of the National Academy of Sciences, the National Academy of Engineering, and the Institute of Medicine. The members of the committee responsible for the report were chosen for their special competences and with regard for appropriate balance.

This study was supported by funding from the Energy Foundation and the George and Cynthia Mitchell Endowment for Sustainability Science. Any opinions, findings, conclusions, or recommendations expressed in this publication are those of the author(s) and do not necessarily reflect the views of the organizations or agencies that provided support for the project.

International Standard Book Number-13: 978-0-309-14714-9
International Standard Book Number-10: 0-309-14714-X

Additional copies of this report are available from the National Academies Press, 500 Fifth Street, N.W., Lockbox 285, Washington, D.C. 20055; (800) 624-6242 or (202) 334-3313 (in the Washington metropolitan area); Internet, *http://www.nap.edu*.

Printed in the United States of America

THE NATIONAL ACADEMIES

Advisers to the Nation on Science, Engineering, and Medicine

The **National Academy of Sciences** is a private, nonprofit, self-perpetuating society of distinguished scholars engaged in scientific and engineering research, dedicated to the furtherance of science and technology and to their use for the general welfare. Upon the authority of the charter granted to it by the Congress in 1863, the Academy has a mandate that requires it to advise the federal government on scientific and technical matters. Dr. Ralph J. Cicerone is president of the National Academy of Sciences.

The **National Academy of Engineering** was established in 1964, under the charter of the National Academy of Sciences, as a parallel organization of outstanding engineers. It is autonomous in its administration and in the selection of its members, sharing with the National Academy of Sciences the responsibility for advising the federal government. The National Academy of Engineering also sponsors engineering programs aimed at meeting national needs, encourages education and research, and recognizes the superior achievements of engineers. Dr. Charles M. Vest is president of the National Academy of Engineering.

The **Institute of Medicine** was established in 1970 by the National Academy of Sciences to secure the services of eminent members of appropriate professions in the examination of policy matters pertaining to the health of the public. The Institute acts under the responsibility given to the National Academy of Sciences by its congressional charter to be an adviser to the federal government and, upon its own initiative, to identify issues of medical care, research, and education. Dr. Harvey V. Fineberg is president of the Institute of Medicine.

The **National Research Council** was organized by the National Academy of Sciences in 1916 to associate the broad community of science and technology with the Academy's purposes of furthering knowledge and advising the federal government. Functioning in accordance with general policies determined by the Academy, the Council has become the principal operating agency of both the National Academy of Sciences and the National Academy of Engineering in providing services to the government, the public, and the scientific and engineering communities. The Council is administered jointly by both Academies and the Institute of Medicine. Dr. Ralph J. Cicerone and Dr. Charles M. Vest are chair and vice chair, respectively, of the National Research Council.

www.national-academies.org

STEERING COMMITTEE ON EXPANDING BIOFUEL PRODUCTION: SUSTAINABILITY AND THE TRANSITION TO ADVANCED BIOFUELS

Patrick Atkins, Aluminum Company of America (ALCOA)
John Carberry (Committee Chair), Former Director, Environmental Technology, DuPont
Peter Ciborowski, Research Scientist, Minnesota Pollution Control Agency
Elisabeth Graffy, Economist, U.S. Geological Survey, Office of the Associate Director for Geography
Nathanael Greene, Senior Policy Analyst, Natural Resources Defense Council
Jason Hill, Research Associate, University of Minnesota
Tracey Holloway, Director, Center for Sustainability and the Global Environment, Assistant Professor, University of Wisconsin-Madison
Marcia Patton-Mallory, Bioenergy and Climate Change Specialist, U.S. Forest Service
Bruce Rodan, U.S. Environmental Protection Agency
Gary Radloff, Director of Policy and Strategic Communications, Wisconsin Department of Agriculture Trade and Consumer Protection

Preface and Acknowledgments

To follow up on discussions held by the Roundtable on Science and Technology for Sustainability, the Science and Technology for Sustainability Program appointed a steering committee of subject matter experts to plan a workshop that would explore further the implications for sustainability of expanding biofuel production. Initial discussions suggested that many local and regional impacts associated with expanding biofuels exist in the U.S. Upper Midwest, so the workshop focused specifically on this region.

In June 2009 the steering committee convened the workshop with the specific purpose of developing a better understanding of the lessons that can be learned from the experience with producing corn-based ethanol and the likely environmental, economic, social, and energy security impacts of advanced biofuels. The workshop offered an opportunity for dialogue between researchers and policy makers on the sustainability impacts of expanding biofuel production at state and regional levels. The workshop also sought to identify policy objectives and challenges facing state officials related to biofuels, provide examples of research that may be useful to state decision-makers, and evaluate various tools and indicators of possible use to state policy makers in assessing the likely sustainability impacts and tradeoffs of their choices.

This document has been prepared by the workshop rapporteurs as a factual summary of what occurred at the workshop. The statements made in this volume are those of the rapporteurs and do not necessarily represent positions of the workshop participants as a whole, the steering committee, the Roundtable on Science and Technology for Sustainability, or the National Academies.

This workshop summary is the result of substantial effort and collaboration among several organizations and individuals. We wish to extend a sincere thanks to each member of the steering committee for his/her contributions in scoping, developing, and carrying out this project.

The project would not have been possible without the financial support of its external sponsor, the Energy Foundation. It also benefitted from internal support provided by the George and Cynthia Mitchell Endowment for Sustainability Science.

This report has been reviewed in draft form by individuals chosen for their diverse perspectives and technical expertise, in accordance with procedures approved by the National Academies' Report Review Committee. The purpose of this independent review is to provide candid and critical comments that will assist the institution in making its published report as sound as possible and to ensure that the report meets institutional standards for quality and objectivity. The review comments and draft manuscript remain confidential to protect the integrity of the process.

We wish to thank the following individuals for their review of this report: Richard Cruse, Iowa State University; Gregory Nemet, University of Wisconsin; Gary Radloff, Wisconsin Department of Agriculture; and Lisa Shames, U.S. Government Accountability Office.

Although the reviewers listed above have provided many constructive comments and suggestions, they were not asked to endorse the content of the report, nor did they see the final draft before its release. Responsibility for the final content of this report rests entirely with the author(s) and the institution.

Patricia Koshel and Kathleen McAllister
Rapporteurs

Contents

I

Introduction and Overview

On June 23 and 24, 2009, the National Research Council's Roundtable on Science and Technology for Sustainability ("Roundtable") hosted the workshop "Expanding Biofuel Production: Sustainability and the Transition to Advanced Biofuels—Lessons from the Upper Midwest for Sustainability" in Madison, Wisconsin. Organized by a steering committee, the workshop was attended by approximately 75 people representing academia, state government, nongovernmental organizations, the business sector, and federal agencies. It was organized around the following topics: policy drivers for the expansion of biofuels; the state of biofuel technologies; the economic, environmental, and social dimensions of sustainability, as related to biofuels; the business of biofuels; tools and indicators for decision makers; and ongoing research related to biofuels and sustainability. Breakout sessions examined lessons learned from the experience with producing corn-based ethanol, the potential impacts of next-generation fuels, and future challenges and opportunities. Throughout the workshop there was substantial discussion about uncertainty—when will next-generation fuels be available at commercial scale; what are the most likely feedstocks and where will they be grown; does ethanol represent the best fuel for the future U.S. transportation system, or are other energy sources, including other bio-based fuels, potentially more sustainable; can policy inconsistencies at both federal and state levels be resolved to support sustainability objectives; how can changes in land use be included as a cost of production; and what are the long term consequences for scarce water resources, ecosystems services, and local communities?

CONTEXT

The U.S. biofuel industry has grown dramatically in recent years, with production expanding from 1.6 billion gallons in 2000 to 9 billion gallons in 2008.[1] This dramatic increase can be attributed to the rise in production of corn-based ethanol and associated, smaller quantities of soy-based biodiesel. The number of refineries has also increased—from 54 in 2,000 to 170 in January 2009.[2] The worldwide economic recession and lower prices for petroleum have slowed the expansion of the industry, but because of strong state and federal mandates, production is expected to grow until production capacity reaches the federally mandated 36 billion gallons of biofuel in 2022.[3]

While energy prices, energy security, and climate change are front and center in the national media, these issues are often framed to the exclusion of the broader issue of sustainability—ensuring that the production and use of biofuels do not compromise the needs of future generations by recognizing the need to protect life-support systems, promote economic growth, and improve societal welfare. Thus, it is important to understand the effects of biofuel production and use on water quality and quantity, soils, wildlife habitat and biodiversity, greenhouse gas emissions, air quality, public health, and the economic viability of rural communities.[4]

Although corn-based ethanol is likely to continue to be a major contributor to U.S. biofuel supply in the near term, it is important to plan for the transition to advanced biofuels, such as agricultural resides (e.g., corn stover), perennial grasses and woody biomass, which are now almost universally viewed as preferable from a sustainability perspective. Decisions have been made at various levels of government to promote biofuels as a potential means of reducing greenhouse gases and enhancing economic development and energy security without a clear understanding of the economic, environmental, and social impacts of biofuel production and use.

While a number of studies have examined some of the environmental impacts associated with the expansion of biofuel production and use, most of these have focused at a national level. For example, the National Academies published a report assessing the water implications of biofuels[5] and the World Resources Institute has also published a series of reports on the subject.[6] However, many

---

[1]See *http://www.ethanolrfa.org/industry/statistics/#A* (accessed July 2, 2009).

[2]See *http://www.ethanolrfa.org/industry/statistics/#EIO* (accessed July 2, 2009).

[3]U.S. Energy Independence and Security Act of 2007 (EISA).

[4]Energy security, while part of the EISA mandate, does not traditionally fall within the scope of sustainability analyses and thus was not part of workshop discussions.

[5]*Water Implications of Biofuels Production in the United States*. NRC 2009, *http://www.nap.edu/catalog.php?record_id=12039*.

[6]*Plants at the Pump: Reviewing Biofuels' Impacts and Policy Recommendations*. World Resources Institute, July 2008; *Biofuels and the Time Value of Carbon: Recommendations for GHG Accounting Protocol*. World Resources Institute, March 2009.

of the environmental effects of corn-based biofuels as well as next generation biofuels are uniquely local or regional—including potential changes in water availability or soil fertility. And many of the economic and social effects are also most pronounced at a local level.

In an effort to better understand these impacts, the steering committee decided to narrow the workshop scope and focus on three states in the Upper Midwest—Iowa, Minnesota, and Wisconsin. This region is undergoing an economic transition from a historical farming and manufacturing economy. Biofuels technology development and increased production have been touted as central to a stronger regional economy. The three states have supported aggressive policies to promote the development of the industry, focused on both the supply side as well as the demand side. In addition, each of these states has strong research universities and a number of academic researchers focused both on the technology aspects of biofuels and on the economic, environmental, and social impacts.

Iowa, Minnesota, and Wisconsin have seen substantial increases in corn production since 2000, with total acreage expanding from 23,000 planted acres in 2000 to 26,650 in 2007, and then declining slightly in 2008.[7] Each state also has a large number of ethanol refineries—39 in Iowa, 17 in Minnesota, and 9 in Wisconsin. These plants account for 35 percent of the total U.S. nameplate capacity.[8] These states are also likely to be an important source of biomass feedstocks for next-generation biofuels. Data from the National Renewable Energy Laboratory suggest that approximately 75,000 tons of biomass resources could be available annually from these three states—almost one-quarter of total U.S. biomass resources.[9]

The workshop was designed to draw on the expertise of researchers and policy makers in the three-state region to better understand these local impacts and the challenges faced by state policy makers, while at the same time recognizing the need to also consider the broader national and global impacts, including impacts on world food supplies.

ORGANIZATION OF THE REPORT

This report is limited in scope to the presentations, workshop discussions, and background documents produced in preparation for the workshop. Chapter 2 discusses the principal policy drivers behind the expansion of biofuel production and use. Chapter 3 focuses on the results of a recent National Academies report

---

[7]National Corn Growers Association. See *ncga.com/corn-production* (accessed July 6, 2009).

[8]See *neo.neb.gov* (accessed July 6, 2009). Name plate capacity is the maximum output of a plant based on conditions designated by the manufacturer. Actual production is likely to be less than this amount.

[9]A. Milbrandt. *A Geographic Perspective on the Current Biomass Resource Availability in the United States*. NREL/TP 560-39181. December 2005. Available at *http://www.nrel.gov/docs/fy06osti/39181.pdf*.

on the status of alternative liquid transportation fuel technologies as well as other efforts to develop alternative transportation fuels. Chapter 4 describes some of the environmental, economic, and social impacts associated with current- and next generation biofuels. Chapter 5 provides a perspective on issues to be addressed as part of the transition to advanced biofuels, including federal policy, research needs, and tools and indicators needed by decision makers to assess the consequences and tradeoffs of expanding production and use.

The report appendixes include the workshop agenda, brief biographies of workshop speakers, a selected bibliography of reports and papers addressing issues of biofuels and sustainability, a background paper describing the biofuels policies in the three Upper Midwest states, and a paper on tools and indicators used to assess various aspects of biofuel production and use. The appendixes also include examples of ongoing federal research programs and projects related to sustainability and biofuels.

II

Policies Driving the Expansion of Biofuel Production

Many presentations at the workshop described increases in the production and use of biofuels over the last decade. These have been driven largely by federal and state policies intended to create a biofuel industry, while at the same time reducing U.S. reliance on imported petroleum, promoting energy security, and decreasing emissions of greenhouse gases (GHGs). These policies include various forms of subsidies as well as mandates for production and use.

FEDERAL LEGISLATION

Key legislative drivers include the Energy Policy Act of 2005 (EPACT), the Energy Independence and Security Act of 2007 (EISA), and the 2002 and 2008 Farm Bills. EPACT set numerical goals for ethanol production—7.5 billion gallons by 2012—and provided credits to refiners and blenders. EISA expanded these mandates, increasing the required production level to 36 billion gallons by 2022 (Figure 1). Of the total, 21 billion gallons are to be obtained from cellulosic and other advanced biofuels.

Energy Independence and Security Act of 2007

EISA's provisions have important implications for the sustainable production and use of biofuels. The act:

- Requires significantly increased volumes of renewable fuel production,

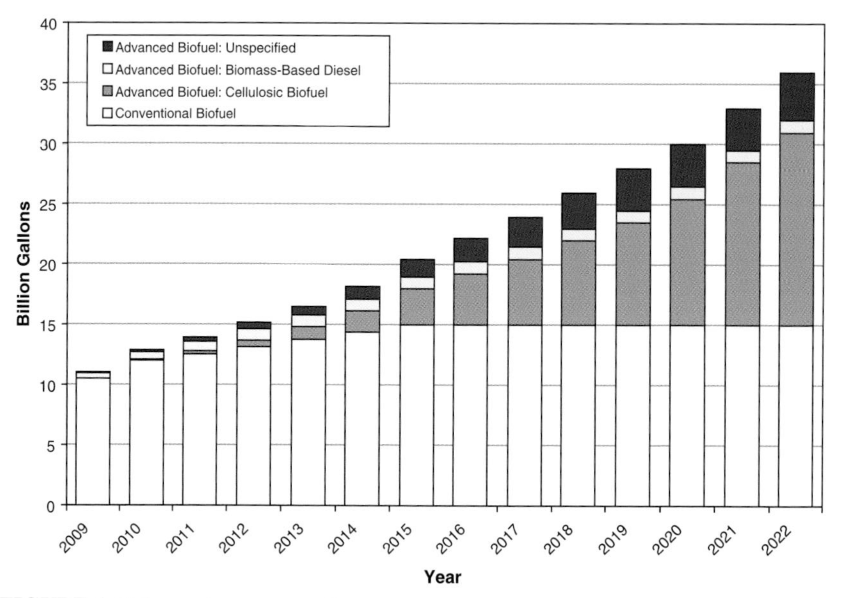

FIGURE 1 Volume changes over time.
Source: U.S. Environmental Protection Agency, Office of Transportation and Air Quality, Workshop Presentation by Bruce Rodan, June 23, 2009.

with separate volume requirements for cellulosic biofuels, biomass-based diesel, advanced biofuels,[1] and total renewable fuels.

- Modifies the definition of renewable fuels to include minimum life-cycle GHG reduction thresholds. These reductions are to include both direct emissions and indirect emissions resulting from significant land-use changes—including international land-use changes.
- Restricts the types of feedstocks that can be used to make renewable fuels and the types of land that can be used to grow feedstocks.
- Includes specific waivers and U.S. Environmental Protection Agency (EPA)-generated credits for cellulosic biofuels.

While EISA has a number of sustainability provisions, it "grandfathers" the first 15 billion gallons/year of biofuel, exempting this amount of fuel from

[1]EISA defines advanced biofuels as renewable fuels, other than ethanol derived from corn starch that has lifecycle greenhouse gas emissions that achieve at least a 50 percent reduction over baseline lifecycle greenhouse gas emissions. Types of advanced biofuels include: ethanol derived from cellulose or lignin, sugar or starch (other than corn starch), or waste material, including crop residue, other vegetative waste material, animal waste, and food waste and yard waste; biomass-based diesel; biogas produced through the conversion of organic matter from renewable biomass; butanol or other alcohols produced through the conversion of organic matter from renewable biomass; and other fuel derived from cellulosic biomass

the EISA's GHG reduction and source requirements. EISA also grandfathers all existing ethanol production facilities, thereby exempting them from meeting the requirements. Only new production, beyond 15 billion gallons/year, must meet the specific GHG requirements outlined in the Act. (See Box 1)

EISA also restricts the types of renewable feedstocks that can be used and the types of lands from which the feedstocks can be derived. For example, feedstocks can be grown on agricultural land that has been cleared and cultivated prior to December 2007, but not on federal land, except for wildfire areas. While there are no other specific environmental requirements, EISA requires EPA, in consultation with the U.S. Department of Agriculture (USDA) and the U.S. Department of Energy, to report every three years on environmental impacts, including:

- Environmental issues, including air quality, effects on hypoxia, pesticides, sediment, nutrient and pathogen levels in waters, acreage and function of waters, and soil environmental quality;
- Resource conservation issues, such as soil conservation, water availability, and ecosystem health and biodiversity, including impacts on forests, grasslands, and wetlands; and
- The growth and use of cultivated invasive or noxious plants and their impacts on the environment and agriculture.

BOX 1
Greenhouse Gas Requirements, EISA 2007

Cellulosic Biofuel: 16 billion gallons by 2022
Renewable fuel produced from cellulose, hemicellulose, or lignin—cellulosic ethanol, biomass-to-liquid diesel, green gasoline, etc. Must meet a 60 percent life-cycle GHG threshold.

Biomass-Based Diesel: 1 billion gallons by 2012 and beyond
e.g., biodiesel, "renewable diesel" if fats and oils are not co-processed with petroleum. Must meet a 50 percent life-cycle GHG threshold.

Advanced Biofuel: Minimum of 4 billion additional gallons by 2022
Essentially anything but corn starch ethanol; includes cellulosic biofuels and biomass-based diesel. Must meet a 50 percent life-cycle GHG threshold.

Renewable Biofuel: Up to 15 billion gallons of other biofuels
Ethanol derived from corn starch, or any other qualifying renewable fuel. Must meet a 20 percent life-cycle GHG threshold; only applies to fuel produced in new facilities.

Source: Energy Independence and Security Act of 2007 (HR6).

Food, Conservation, and Energy Act of 2008

In addition to EISA, the Food, Conservation, and Energy Act of 2008 (the 2008 Farm Bill) has a number of provisions encouraging the expansion of biofuel production and use, including tax credits for ethanol, blender credits for cellulosic fuels, and continuation of import duties on imported ethanol. One of the Farm Bill's most important provisions is USDA's Biomass Crop Assistance Program, which provides payments to farmers for growing new feedstocks and subsidizes the costs of collection, harvest, storage, and transportation to conversion facilities.[2]

STATE POLICY INCENTIVES

Iowa, Minnesota, and Wisconsin have also developed a set of policy incentives to encourage development of a local biofuel industry.[3] During the workshop, state representatives and researchers described current and planned state biofuel policies.

Wisconsin

Wisconsin uses a combination of financial and regulatory incentives to encourage industry development—making the state a "market participant" in an industry promoted heavily through federal government regulation. For example, the state's Ethanol and Biodiesel Fuel Pump Income Tax Credit allocates 25 percent (or up to up to $5,000) of the cost of installation for ethanol and biodiesel purveyors. Wisconsin has also proposed an income tax credit for biodiesel production—10 cents per gallon, with a minimum production of 2.5 million gallons and a maximum credit of $1 million. Laws were also passed mandating that state employees operate flex-fuel vehicles whenever possible and use alternative fuels, as Wisconsin is attempting to reduce its petroleum consumption by 20 percent by 2010 and 50 percent by 2015. However, the current lack of E85[4] facilities proves to be a significant challenge for the industry.

Wisconsin's Department of Commerce has also established an Energy Independence Fund, whereby the governor has committed $150 million over 10 years, encouraging energy independence. Thus far, $22.5 million has been awarded—mainly to R&D projects on advanced biofuels and for additional research on improving the efficiency of existing biofuel plants. However, due to budget cuts, this program is suspended until 2011. Although Wisconsin continues to promote the state biofuel industry through various incentive programs, the current eco-

[2]See *http://www.fsa.usda.gov/FSA/webapp?area=home&subject=ener&topic=bcap*.

[3]Note no formal presentation was made about Iowa's biofuel policy. Information on Iowa programs, however, is included in Appendix D.

[4]E85 is a fuel blend of 85 percent ethanol and 15 percent gasoline.

nomic downturn and the uncertainty of the market have forced many ethanol plants to be idle.

Minnesota

Minnesota was the first state to develop an ethanol mandate requiring that all gasoline sold in the state contain 10 percent ethanol, increasing to 20 percent by 2013. The state also created a variety of biofuel incentives—blenders' credits, producer payments, tax benefits for refineries under the state's Job Opportunity Building Zone (JOBZ) Program, reduced fuel taxes for consumption of E85, and grants for the development of next-generation fuels. For older plants, blenders' credits for ethanol were issued through a producer payment program for ethanol plants built before 2000—issuing a credit for 20 cents per gallon of ethanol produced, up to 15 million gallons of ethanol per year per plant. Newer ethanol plants are covered by JOBZ, which is a more general economic development program (i.e., not solely a biofuel industry program) that provides financial incentives and tax credits/breaks to a variety of businesses.

By 2015, one-quarter of Minnesota ethanol supplies must come from cellulosic feedstocks. Also, Minnesota was the first state to institute a biodiesel mandate—currently 5 percent and increasing to 10 percent in 2012 and 20 percent in 2015. However, like Wisconsin, Minnesota's biofuel industry has suffered during the current economic decline, and many of the state's larger plants have been shut down. Meeting the 5 percent target as well as the latter goals will be difficult unless the industry can recover economically.

Recent scientific data and pressure on declining state budgets have to some extent eroded support for biofuels in Minnesota, leading the state legislature to commission an analysis of the scientific literature and the specific impacts of state subsidy policies. The legislative auditor's report[5] concluded that traditional corn-based ethanol and soy-based biodiesel have reduced petroleum consumption and have provided some economic development benefits in rural areas, while also causing some negative environmental impacts. Some of these impacts—especially increases in nitrous oxide emissions and the effects of changes in land use and water availability—have not been fully assessed, but are in need of critical analysis as the industry expands. Where the biomass would be grown was also raised as one of the report's critical issues, as well as the associated land-use and environmental impacts. The report also questioned the need for state subsidies, noting that they now account for a very small percentage of producer revenues and are unlikely to play a major factor in business decisions. The report concluded that if Minnesota intends to scale up its biofuel industry to meet the goal of increasing cellulosic biofuel production, additional studies

[5]Office of the Legislative Auditor, State of Minnesota. *Evaluation Report—Biofuel Policies and Programs*. St. Paul, Minnesota. April 2009. Available at *http://www.auditor.leg.state.mn.us/PED/pedrep/biofuels.pdf*.

must be conducted to mitigate negative environmental and economic impacts. The report also strongly encouraged the Minnesota state legislature to remove the subsidies and credits for older ethanol plants, citing rising profits for plants that still receive the subsidies.

EISA grandfathers existing production facilities thereby providing no incentive to improve production practices or increase efficiency. New production facilities will be required to reduce by at least 20 percent the life cycle greenhouse gas (GHG) emissions relative to life cycle emissions from gasoline and diesel. Biorefineries will qualify for cash awards for producing fuels that displace more than 80 percent of the fossil-derived processing fuels used to operate a biorefinery. Workshop participants raised a number of concerns about current policies and the lack of incentives for performance improvements and innovation. In particular, many participants suggested that the current policy framework sends mixed signals to producers and consumers. For example, EISA grandfathers existing production facilities, thereby discouraging efficiency improvements in these facilities. Current policies effectively reduce the cost of biofuels, encouraging greater consumption rather than the development of more fuel-efficient vehicles. And policies do not provide adequate means of fully accounting for the potential loss of ecosystem services caused by increasing soil erosion, water use, etc.

New climate legislation, which was being debated in Congress during the workshop, was seen as potentially exacerbating potential negative land-use and environmental costs and diluting the positive environmental provisions of previously enacted legislation. Decisions to delay provisions allowing for the calculation of indirect land-use impacts under EPA's new renewable fuels standard and the potential for expanding feedstock production on environmentally sensitive lands were particularly troublesome to many participants, as were decisions to shift some responsibilities for administering EISA from EPA to USDA.

State representatives at the workshop implied that they were waiting for federal leadership before proposing new energy policies and expressed frustration with the complexity and slow-moving federal policy process. They suggested that a federal framework with clear goals and metrics was needed to address climate change and to support the development of a sustainable domestic biofuel industry. While the state representatives recognized the role of the states in supporting both biofuel and climate goals, they expressed frustration with conflicting federal energy policies.

III

Next-Generation Technologies and Feedstocks

Several presenters described the efforts of federal agencies and the private sector to develop next-generation bioenergy technologies and prospects for transitioning from a biofuel industry dominated by corn-based ethanol to one based on a more diverse set of feedstocks. One of the largest of these programs is the U.S. Department of Energy's (DOE's) Biomass Program.[1] This program is currently focused on deploying cellulosic technologies—building pilot commercial-scale biorefineries, often partnering with industry. The program also conducts basic technology development research focused both on cellulosic ethanol as well as on other advanced fuels, such as green diesel and green gasoline, which can be substituted for petroleum-based fuels.

Annual DOE funding for these activities has averaged about $100 million. The 2009 stimulus funding increased the level of funding dramatically—by an additional $800 million. The additional funds are being used for demonstration and pilot-scale refineries, as well as supplementing previously funded commercial-scale biorefinery projects. DOE also funds analytical work in the areas of life-cycle analysis of water, greenhouse gas emissions, and land-use changes.

DOE currently funds three Bioenergy Centers, one which includes a focus on sustainability, the Great Lakes Bioenergy Research Center (GLBRC) in Madison, Wisconsin. The GLBRC sustainability program is designed to improve carbon balances across the entire biofuel life cycle and to seek ways to enhance ecosystem services in biofuel landscapes. Other GLBRC activities seek to improve plant biomass, biomass processing, and cellulosic conversion technologies.

Many private companies, such as British Petroleum (BP), are also conduct-

[1]See *http://www1.eere.energy.gov/biomass/*.

ing research on next-generation biofuels. In addition to supporting the Energy Biosciences Institute in Berkeley, California, BP has formed partnerships with DuPont and Verenium.

- The DuPont program is focused on developing efficient ways to produce biobutanol, a fuel with a lower emissions profile and higher energy density than corn-based ethanol. A pilot plant is under construction in the United Kingdom, and a second plant is expected to be built in the United States in the 2012-2013 timeframe.
- BP is also partnering with Verenium, a startup company developing cellulosic conversion technologies. It is planning to build the first cellulosic commercial-scale biofuel production plant in Florida next year with full production predicted to begin by 2012. The plant will use a biochemical pathway that BP expects to be more competitive in the long run because costs are not as dependent on scale as are plants using thermal chemical or biochemical processes.

During the workshop, a representative from a venture capital firm talked about research being done by ZeaChem, which bypasses more traditional thermochemical and biochemical processes. The new process can be used to produce both biofuels and industrial chemicals using cellulosic feedstocks.

Another presenter described efforts to develop other biomass-derived fuels—hydrocarbon biofuels. He explained that hydrocarbon biofuels have the same energy content as petroleum, and thus do not create a mileage penalty. He added that these fuels can use the existing infrastructure facilities developed for gasoline—transport pipelines, fuel pumps, and storage facilities eliminating the need to duplicate infrastructure.

Several presentations discussed potential future feedstocks. For example, the U.S. Forest Service and the Oak Ridge National Laboratory are currently updating bioenergy feedstock estimates in the 2005 billion-ton study.[2] The initial study suggested that about 400 million tons could be provided from wood sources—logging residue, forest thinnings, mill residue, and urban wood wastes. Short-rotation woody crops were counted as an agricultural source. These estimates are now being revised to indicate the economic feasibility and sustainability of woody biomass feedstocks at a county level. Unlike other potential cellulosic feedstocks, woody biomass already represents a large share of total U.S. renewable energy supplies, and can be used for liquid fuels as well as to produce electricity and heat. With more than half the states in the nation now having renewable portfolio

[2]Perlack, Robert D., Lynn L. Wright, and Anthony F. Turhollow. 2005. *Biomass as a Feedback for a Bioenergy and Bioproducts Industry: The Technical Feasibility of a Billion-Ton Annual Supply.* A report prepared for the United States Department of Energy and the United States Department of Agriculture. Oak Ridge, TN: Oak Ridge National Laboratory. Available at *www.ornl.gov/~webworks/cppr/y2001/rpt/123021.pdf.*

standards, demand for woody biomass to produce electricity is likely to grow, competing with its use as a liquid transportation fuel feedstock.

Several federal agencies and land grant universities are collaborating in a regional biomass feedstock partnership to conduct field trials of potential feedstocks and to assess the impacts of these crops on soil carbon, hydrology, and water quality, as well as direct greenhouse gas emissions. An important aspect of the research is exploring how energy crops can best be integrated with current cropping systems. USDA is supporting research to assess how crop residues, such as corn stover, can be used as cellulosic feedstocks and harvested in ways that maintain soil organic carbon and protect croplands from erosion. USDA is also conducting research to develop varieties of perennial grasses and management practices that promote greater biomass feedstock yields.

One important issue that arose in numerous discussions was the need to understand the impacts of changes in land use. Many participants expressed concern about potential negative impacts associated with the expansion of biofuel production on marginal lands and the withdrawal of land from the Conservation Reserve Program (CRP). Growing economic pressures are likely to lead to the expansion of feedstock production on these lands without an adequate understanding of the value of the ecosystem services provided by these lands and the potential impacts if these services are lost. However, some research is now underway to assess the likely impacts of changes in land use associated with the expansion of energy crops and the potential effects on watershed scale hydrological flows, changes in soil nutrients, biodiversity, and pest suppression. (Appendix F).

A member of a National Academies committee assessing the status of various technologies for the production of alternative liquid transportation fuels discussed the major conclusions of a recent report.[3] The study found that biomass (from plants and wastes) could be cost-competitive with petroleum over the next 10-25 years, leading to lower greenhouse gas emissions and reduced dependence on imported petroleum. The report estimates that approximately 500 million tons of biomass feedstocks could reasonably be produced annually and converted to fuels without major environmental impacts or impacts on food availability (Table 1). Different cellulosic feedstocks with woody biomass are expected to have the lowest costs, followed by straw and high-yield grasses. The report suggests that 0.5 million barrels/day of gasoline equivalent can be produced by 2020 and 1.7 million barrels/day by 2035.

Reaching these levels by 2020 will require increased funding for large demonstration facilities and adoption of low-carbon fuel standards; a carbon price, or explicit carbon-reduction targets; and accelerated federal investments in these

[3]America's Energy Future Panel on Alternative Liquid Transportation Fuels, National Academy of Sciences, National Academy of Engineering, and National Research Council. 2009. *Liquid Transportation Fuels from Coal and Biomass: Technological Status, Costs, and Environmental Impacts.* Washington, DC: National Academies Press. Available at *http://www.nap.edu/catalog.php?record_id=12620.*

TABLE 1 Estimated Cellulosic Feedstock Production for Biofuels

	Millions of Dry Tons	
Feedstock Type	Current	2020
Corn stover	76	112
Wheat and grass straw	15	18
Hay	15	18
Dedicated fuel crops	104	164
Woody residues[a]	110	124
Animal manure	6	12
Municipal solid waste	90	100
Total	416	548

[a]Woody residues currently used for electricity generation are not included in this estimate.

Source: NRC America's Energy Futures Report: "Liquid Transportation Fuels from Coal and Biomass: Technological Status, Costs, and Environmental Impacts," Workshop Presentation by John Miranowski, June 23, 2009.

new technologies. Many participants noted that to ensure the sustainability of these new fuels, economic incentives will also need to be provided to farmers and developers to use a systems approach—addressing soil, water, and air quality; carbon sequestration; wildlife habitat; and rural development. As it is expected to take at least until 2030 to attain large-scale cellulosic fuel production, most participants agreed that meeting this goal will require the building of tens to hundreds of conversion plants, as well as associated transport and distribution infrastructure facilities.

IV

Dimensions of Sustainability and Expanding Biofuel Production

This chapter summarizes workshop presentations and discussions that focused on defining what sustainability means in the context of biofuel production and more broadly transportation systems. It describes some of the likely environmental, economic, and social impacts associated with the expanded production and use of both corn-based ethanol and next-generation biofuels.

To provide a context for examining the sustainability dimensions of biofuels, different definitions of sustainability were discussed. The most widely used definition is derived from the Brundtland Commission report, *Our Common Future*: "Sustainability meets the needs of the present without compromising the ability of future generations to meet their own needs."[1] While this definition was seen as useful conceptually, it was not seen as a practical construct for policy makers. The biologist E.O. Wilson offered an alternative definition: "The common aim must be to expand resources and improve quality of life for as many people as heedless population growth forces upon Earth, and do it with minimal prosthetic dependence. That, in essence, is the ethic of sustainable development."[2] This implies the need for decision makers to consider the ethical implications surrounding a problem or issue—such as potential tradeoffs between food production and fuels—as well as the need to apply a broad systems perspective.

Life-cycle assessment (LCA) is often used to evaluate the sustainability of biofuels from a systems perspective. However, as shown in Figure 2, "attributional" LCA analyses do not address economic or social impacts, and generally focus only on the directly attributed environmental impacts.

---

[1]World Commission on Environment and Development. 1987. *Our Common Future.* Oxford University Press. Available at *http://www.un-documents.net/wced-ocf.htm.*

[2]E.O. Wilson. 1998. *Consilience: The Unity of Knowledge.* New York: Alfred A. Knopf, Inc.

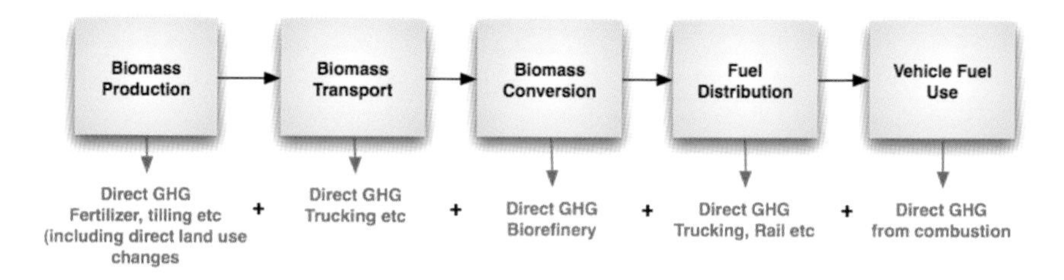

FIGURE 2 "Attributional" LCA, a systems view.
Source: Presentation by John Sheehan, University of Minnesota, June 24, 2009.

The workshop presenter suggested that a more appropriate approach would be to consider a "consequential" LCA, which could consider both the immediate or direct impacts as well as the indirect impacts, although still not fully assessing the economic or social impacts (Figure 3).

While these analytical tools still do not provide clear guidance for policy makers or investors, they do create a tool for dialogue. A number of workshop participants suggested that more comprehensive systems frameworks are needed to examine the interconnected environmental, economic, and social impacts and to allow the outcomes of alternative systems to be consistently evaluated and

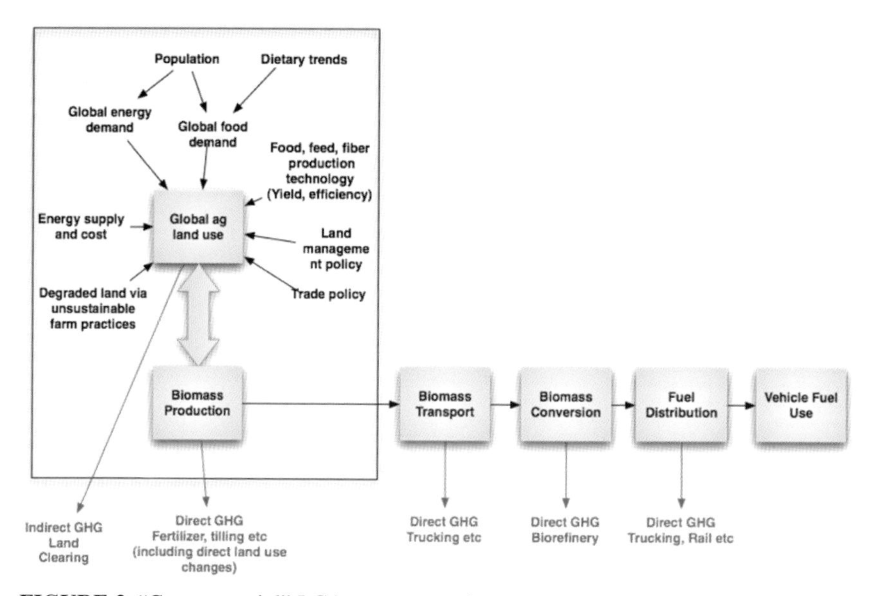

FIGURE 3 "Consequential" LCA, a systems view.
Source: Presentation by John Sheehan, University of Minnesota, June 24, 2009.

compared. A frequent theme throughout the workshop was the need to have tools that would allow decision makers to consider tradeoffs between various feedstocks, conversion technologies, feedstock sources, location of refineries, the characteristics and conditions of local environmental resources, and the environmental, health, economic, and social impacts at various scales.

Other tools mentioned included "standards" or certification schemes, many of which include social and economic effects. A number of domestic and international organizations are in the process of developing these standards, including the Council on Sustainable Biofuel Production, the Global Bioenergy Partnership, and the Roundtable on Sustainable Biofuels. (These activities are described in the background paper in Appendix E.)

ECONOMIC IMPACTS

The workshop's discussion on the economics of biofuels focused both on the business side of the biofuel industry and on its economic impacts—how the industry has changed local and regional job markets, prices, and government budgets. Many of the "policy drivers" that led to the expansion of the biofuel industry were first put in place to create rural economic development opportunities, boost the price of corn by fostering an industry based on corn, and reduce U.S. dependence on imported petroleum. In addition to these policies, two important events accelerated the growth of the industry—state decisions to phase out the use of methyl tertiary butyl ether (MTBE) as a fuel oxygenate and Hurricane Katrina. The lesson from Hurricane Katrina was how incredibly vulnerable our energy system is in the United States, especially when refinery capacity is primarily located in Gulf States or prime hurricane path routes.

MTBE has been banned in most states because of concerns about groundwater contamination. In 2005, EPA refused to grant liability protection to manufacturers of MTBE, forcing a search for substitutes. Ethanol turned out to be a good substitute, and a market was found for increased ethanol production. Hurricane Katrina not only sharply reduced U.S. petroleum refinery capacity; it also made it difficult to export corn, increasing supplies in the Midwest, and drove down prices. These lower prices, along with federal and state incentives, helped drive the rapid expansion of the biofuel industry. From 2000 to 2008, production increased from 1,630 million to 9,000 million gallons and the number of refineries increased from 54 to 139 with approximately 61 refineries under construction during the period (Figure 4). Investors flocked to a proven technology using a traditional agricultural commodity as a feedstock, and early investors were able to quickly recoup their initial investments—often in less than a year.

The economics of the industry began to shift in late 2008 when the price of petroleum began to fall, corn prices remained high, and the overall U.S. economy began to decline, stifling demand. Formerly profitable refineries were no longer profitable and overall profit margins declined (Figure 5). Plans for

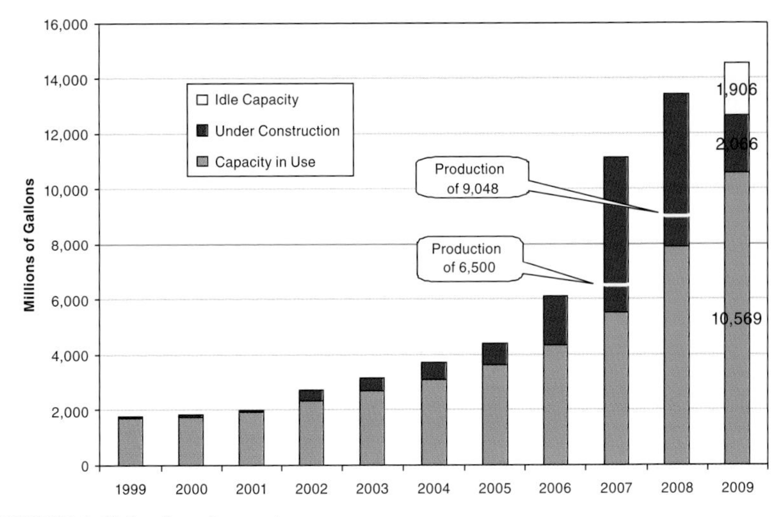

FIGURE 4 U.S. ethanol capacity.
Source: Workshop Presentation by Doug Tiffany, University of Minnesota, June 24, 2009.

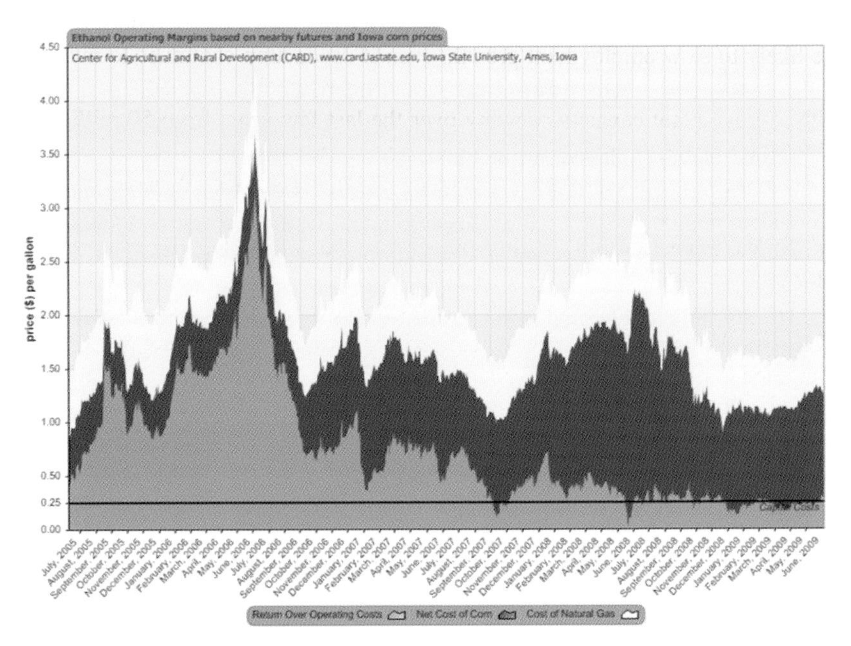

FIGURE 5 Misfortune: Collapsing margins.
Source: Workshop Presentation by David Swenson, Iowa State University, June 24, 2009.

building new refineries were put on hold; at least one major refinery owner—Vera Sun—declared bankruptcy, closing 12 plants with 1.2 million gallons of annual capacity; and another 11 refinery operations also closed.

One of the key objectives of both federal and state biofuel policies was to create "jobs for rural America." Hundreds of thousands of new jobs were promised. In fact, many jobs have been created, but far fewer than originally promised or as claimed by the industry's vocal spokespeople. In 2008, the Renewable Fuels Association claimed that almost 500,000 jobs had been created by the industry. In contrast, data from the U.S. Commerce Department for the same period show only 7,000 ethyl alcohol production jobs. While this 7,000 figure clearly does not reflect all the jobs created by the industry, it is highly unlikely that the multiplier would be 100.[3]

Data from the Iowa Renewable Fuels Association (IFRA) show that more than 83,000 jobs were created by the state's ethanol industry in 2008—almost 40,000 more than it claimed in 2007. Information presented at the workshop by Dave Swenson of Iowa State University suggests that these numbers dramatically overstate the job creation impacts of the industry by counting farm workers who were already engaged in growing corn (30,000 of the 83,000), and counting construction workers engaged on a short-term or temporary basis. The IRFA numbers also appear to exaggerate the number of jobs created indirectly though industry suppliers and jobs created by increased household spending.

Continued improvements in plant efficiency and realized economies of scale are likely to slow employment growth, even if production continues to increase in the future. The industry has already realized increasing economies of scale, with average plant capacity growing over the last few years from 50 million to 100 million gallons a year. Furthermore, process changes and greater economies of scale have increased plant efficiencies and reduced labor demand per unit of output. The average job creation impact of a 50-million-gallon-a-year plant was shown to be 133 jobs, while a plant twice that size produced only 36 more jobs.[4]

Despite these real job gains, the industry has not turned around the loss of rural jobs. In Iowa, farm employment and the number of farm proprietors have continued to decrease, and rural counties have continued to experience population decline—more than 45,000 between 2000 and 2007.[5]

The dramatic expansion in the ethanol industry had major effects on a variety of prices ranging from food and feed and agricultural land to gasoline. The effects of biofuel production on domestic and international food prices were raised as an

[3]U.S. Census Bureau, 2007 Economic Census, *http://www.census.gov/econ/census07/www/*using_*american_factfinder/*, Last accessed: December 29, 2009.

[4]Swenson, D. 2007 Estimating the Future Economic Impact of Corn Ethanol Production in the U.S., Iowa State University.

[5]Iowa, State and County Census Facts, U.S. Census Bureau State and County Quick Facts, *http://quickfacts.census.gov/qfd/states/19000.html*.

ethical concern by a number of workshop participants. They acknowledged that food price increases have been driven only in part by the expansion of biofuel production increases in the United States, as well as in other parts of the globe. Domestically, expanded biofuel production was linked to increases in corn prices, leading to higher feed costs and increasing prices for meat and dairy products. Many participants acknowledged that a number of other factors were linked to increasing food prices—rising petroleum prices, increasing food demands driven by population growth, increasing per-capita consumption levels, the dollar devaluation, and general increases in production costs. Nonetheless, participants believed that there were critical tradeoffs between land used for food and land used for feedstocks for fuel. The projected expansion of biofuel production—whether cellulosic or corn based—will directly and indirectly affect land use.

The price of agricultural land rose sharply over the last few years, in part because of the increasing demand for corn and the promise of ever-increasing farm revenues. In Iowa, the price of agricultural land increased by more than 100 percent between 2000 and 2007.[6] Data for 2008 show some slowing in the growth of farmland values, presumably tied in part to the declining fortunes of the ethanol industry.

While ethanol represents less than 3 percent of U.S. transportation fuels, its production has had a significant effect on retail gasoline prices. Information presented at the workshop suggests that ethanol production has led to relatively large reductions in overall gasoline prices, in part, by creating more domestic refining capacity. The availability of somewhat lower-priced gasoline has increased overall demand. Many participants noted that if gasoline consumption continues to grow faster than ethanol production, there will be no reduction in the nation's need to import petroleum, making it yet more difficult to achieve energy independence—one of the principal objectives of the U.S. biofuels policy.

Bruce Babcock, of Iowa State University, stated that the price of petroleum is critical to determining profits for the biofuel industry. Since ethanol is a substitute for petroleum, it closely tracks the price of oil. This makes the industry very vulnerable to volatile petroleum markets, as was evident during 2008. As petroleum prices dropped sharply, the profit margins of refinery operators fell precipitously. Farmers who thought ethanol production would serve as a hedge against declines in commodity prices have been disappointed. They assumed that during periods of low corn prices they would make large profits from ethanol refineries, and that when corn prices were high they could make money by selling corn for food and feed. However, the price of ethanol is not correlated with corn prices. Corn and ethanol prices can both be low, cutting or eliminating profit margins.

Incentives in the form of tax credits, tax rebates, and various forms of subsidies enacted by both the federal government and many state governments have

[6]Iowa Land Value Survey, Iowa State University, University Extension *www.extension.iastate.edu/landvalues*, Last Accessed December 29, 2009.

been costly. Estimates suggest that the overall cost of these incentives is as high as $8 billion-$11 billion[7] a year, and can be expected to increase as the provisions of the EISA and the 2007 Farm Bill come into play, and more attention is focused on promoting the development of a cellulosic-based industry. Incentive programs promoted in the Upper Midwest have been costly and are now coming under increasing scrutiny because of the current state budget problems, questions about their effectiveness, and uncertainty about federal energy and climate policy. Many participants noted that less obvious are the costs to states and local communities to expand and maintain the transportation infrastructure necessary to move increasing volumes of feedstocks and biofuels to intended users, as well as the need to pay for new supplies for first responders in the event of ethanol fires.

Economics and Next-Generation Fuels

EISA mandates dramatic increases in the use and production of renewable fuels. Overall levels are to increase production from 9 million gallons in 2008 to 36 million gallons in 2022, with the increase after 2016 in advanced biofuels—primarily cellulosic ethanol. This means that in the first years—2010-2012—the cellulosic industry must grow by more than 100 percent a year. Even during 2020-2022, the industry is projected to grow by more than 20 percent a year. Bruce Babcock of Iowa State University noted that no U.S. industry has ever grown that fast. While the corn-based ethanol industry's expansion was dramatic, the year-to-year increase was only 25-30 percent at its highest.

Participants almost universally said this rate of expansion is unlikely because the technology is not yet available on a scale that would sustain this growth. It is unclear which feedstocks or combination of feedstocks are going to be most viable and what they will cost. New production will need to compete with corn-based ethanol—a proven technology and feedstock with far less technical and operational risks. And it is likely that improvements in efficiency will continue driving down the costs of corn-based ethanol. Dramatic increases in cellulosic ethanol production will require enormous new capital, estimated by one presenter to be over $60 billion. Based on the required level of investment and recent experience with corn-based ethanol, investors see significant business risks—far more than was the case with first-generation ethanol. For the foreseeable future, the credit markets are expected to remain tight and venture capital funding will continue to be scarce. Many of the technology uncertainties have been covered earlier, so this section will examine some of the other economic barriers facing potential investors.

Investors are looking for ways to minimize risk and maximize returns. The business case for advanced biofuels depends on a variety of factors on both the supply side and the demand side. The federal government and private investors

[7]D. Koplow "Biofuels in the Midwest—A Discussion," *www.wilsoncenter.org*.

are supporting research to allow for the commercialization of advanced biofuels. The new economic stimulus plan includes almost $800 million for biofuels research in addition to funds allocated in the fiscal year 2009 budget of more than $200 million. And many experiments are being conducted assessing potential feedstocks. However, the returns to investment in advanced biofuels are highly uncertain, in part because promises of low-cost feedstocks grown on marginal land have not been confirmed or analyzed comprehensively to determine the unintended consequences associated with these feedstocks. Investors are looking for consistent supplies and low-cost feedstocks. To some extent the provisions of EISA and EPACT and evolving federal and state renewable fuel standards provide some assurances that there will, in fact, be a demand for both corn-based ethanol and advanced biofuels and create a floor price.

Bruce Babcock of Iowa State University described how the renewable fuel standards (RFS) in both EPACT (RFS 1) and EISA (RFS 2) support investors. RFS 1 effectively provides a guaranteed market for investment that has already taken place. The mechanism for enforcing the standard is the renewable identification number (RIN), which is equal to the RFS—the mandated level of biofuel use. During a year, when companies choose to purchase biofuels, they receive the RIN associated with that purchase. If they do not choose to purchase biofuels, then they can purchase the RIN instead and meet the RFS mandated level. If the demand for biofuel is low, they will start purchasing the RIN, but when they enter the RIN market, the price of RIN will begin to rise reflecting the increase in demand. As the price of the RIN rises, because each gallon of biofuel includes a more valuable RIN with it, the price of biofuels will begin to rise, and biofuel production facilities will re-open because their product's value is rising. In early 2009, the price of gasoline fell so low that no one wanted to buy the more expensive biofuels. Then the price of the RIN started to increase until the price of ethanol increased, which led to the re-opening of many ethanol production facilities—enabling the RFS to be met.

The price of the RIN will only rise enough to keep the least-efficient production plants running in order to meet the mandates of the RFS. The more efficient plants will stay in operation, but as the price of ethanol rises, the less efficient plants will begin to come on line. The price of the RIN not only covers the operational cost of feedstock production, but also accounts for the labor costs and the cost of natural gas. The RFS will help to cover operational costs but will not provide a return on investment, therefore doing nothing to stimulate new investment.

Babcock explained that the RFS 2 makes things yet thornier for investors because it includes "waivable mandates" that allow the EPA Administrator to change the level of the biofuel production mandate. Basically, if the plants are not built, no capacity exists to meet the mandate, and the mandate must be waived—effectively eliminating any incentive for early investors. The price of the RIN with a waivable mandate is only going to cover operational and not capital costs. Therefore, additional tax credits or subsidies will still be needed to induce investment.

Other barriers are also hampering the widespread use of expanded biofuel supplies. Flex-fuel vehicles have been widely promoted, but they still represent a very small portion of the total vehicle stock. And while these vehicles are engineered to operate on a variety of fuel blends up to E85, currently few distribution outlets are selling E85, so many flex-fuel vehicles use standard gasoline. For most of the vehicle stock, EPA regulations limit fuel blends to 10 percent ethanol. While there have been some attempts to increase this level, to date EPA has not changed its regulations and is continuing to test the effect of higher blends on engines and tailpipes. This "blend wall" effectively limits demand.[8]

Participants discussed prospects for a number of other bio-based fuels that would not depend on new storage and distribution infrastructure, such as biobutanol and "green gasoline." In fact, the lack of adequate distribution and storage facilities was cited as a major barrier to the expansion of the biofuel industry. At the time of the workshop, neither the federal government nor private investors were creating the necessary infrastructure.[9]

In addition, some participants cautioned that too much attention may be focused on biofuels, when there are other ways to increase America's energy independence and reduce the growth of greenhouse gas emissions, such as increased fuel efficiency and plug-in hybrid vehicles.

Several participants suggested that future biofuel production should meet the following objectives: reduce land-use pressures and greenhouse gas emissions, use non-food feedstocks, and compete with fossil fuels without subsides. They suggested that a price or tax on carbon would promote more efficient biofuel production.

ENVIRONMENTAL IMPACTS

The major environmental issues associated with expanding biofuel production are greenhouse gas emissions, land use, water use, air and water quality, biodiversity, and human health. Currently most biofuel production relies on corn or soybeans as feedstocks. These are annual crops requiring significant water inputs, including water for irrigation in some regions, as well as fertilizers and pesticides. The negative impacts associated with corn-based ethanol have been widely reported (see the Selected Bibliography in Appendix F). Recent studies suggest that improved corn yields and more efficient refineries improved the en-

[8]A decision, by the U.S. Environmental Protection Agency (U.S. EPA), to raise the blend wall limit will not be made until mid 2010—pending further research on impacts of increasing the blend wall.

[9]However, in the months following the workshop, DOE began a major deployment of infrastructure development programs.

vironmental performance of corn-based ethanol, but despite these improvements cellulosic-derived fuels are thought to be more sustainable.[10]

One presenter defined a sustainable biofuel system as one that is carbon negative with respect to climate, is nutrient and water conservative, provides biodiversity benefits, and has a positive impact on human health. He noted that the promise of advanced biofuels is based on their perenniality and crop diversity—versus annual corn, the feedstock currently used to produce most ethanol. For example, annual cropping systems have a nitrous oxide footprint four to five times greater than that of cellulosic crops and deplete levels of soil carbon. More diverse landscapes also increase levels of ecosystem services, including biocontrol services and cleaner water. However, not all cellulosic systems are created equal. The extent to which this promise is realized will depend on:

- The choice of crops (e.g., annual versus perennial, native versus exotic, invasive versus non-invasive, landscape diversity);
- Management practices (e.g., residue return, harvest timing and intensity, fertilization rate, irrigation); and
- Location (e.g., What crops have been raised before? Whether energy crops will be grown on land previously enrolled in the CRP).

Even with advanced biofuel feedstocks, however, the environmental benefits may be difficult to fully realize. For example, crop residues, such as corn stover, are often cited as a promising cellulosic feedstock. However, if the removal of these residues from fields is not managed effectively, the loss of these field residues could increase soil erosion and nutrient loss and cause soil water loss. Local soil temperatures could rise—creating localized climate effects and overshadowing global warming benefits.

The water impacts of expanding biofuel production, primarily corn-based ethanol, were cited by a number of participants as a major long-term problem for the biofuel industry—a problem that was likely to become more of a constraint with climate change. Water consumed during crop cultivation is significantly more than that consumed by fuel processing facilities, though data monitoring to fully assess water demands is difficult. Current ethanol processing requires approximately 3 gallons of water for every gallon of fuel produced. Only limited data exists for the water resource requirements for cellulosic and algae feedstock production and fuel processing.[11] While some of this water may be recovered, its negative impact on aquifers and other water resources remain a serious local issue.

The increased use of nitrogen-based fertilizers to improve corn yields has led to large amounts of leaching, with only 40 percent of the nitrogen actually going

[10]U.S. General Accountability Office, 2009, Biofuels: Potential Effects and Challenges of Required Increases in Production and Use (GAO-09-446).

[11]Issues regarding algae feedstock production were not thoroughly discussed at this workshop.

to the plants. For example, in the Midwest, the excess nitrogen is deposited into water bodies and eventually travels to the Gulf of Mexico. The excess nitrogen in the Gulf causes large algal blooms that decompose, using up oxygen and creating a hypoxic zone. This zone has increased significantly in recent years, and is likely to continue to expand with projected increases in exports of dissolved inorganic nitrogen, despite pledges in 2005 to address its root causes.

There are also concerns about local groundwater quality as evidenced by increased nitrate-nitrogen concentrations. A number of wells in Wisconsin's Dane County now exceed recommended EPA levels of 10 parts per million.

The health and safety impacts, both positive and negative, of biofuel production and use have received only limited attention with most studies on corn based ethanol or soybased biodiesel not advanced biofuels. Understanding and mitigating potentially significant negative impacts are critical to evaluating future renewable fuel options. There are recognized health and safety impacts along the entire biofuel supply chain—beginning with feedstock production and moving progressively through feedstock logistics, biofuel production, biofuel distribution, and biofuel end use (Figure 6). There are also likely to be indirect effects on human health. The scope of these impacts will depend on the types of fuel, feedstock, and conversion technologies and the characteristics of individual places (e.g., population density and baseline measures of air and water quality).

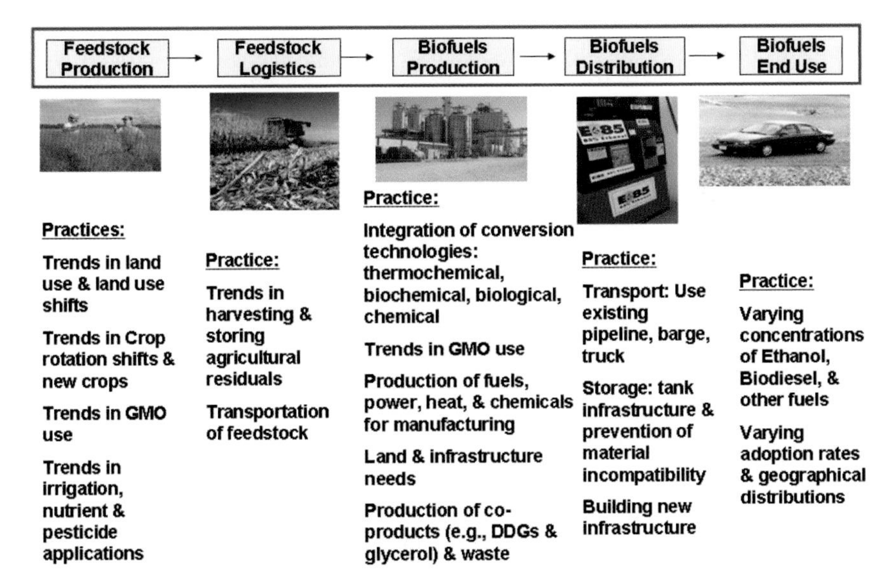

FIGURE 6 Practices and technologies of the biofuels supply chain.
Source: Workshop Presentation by Donna Perla, U.S. Environmental Protection Agency, June 24, 2009.

Some of these implications include the following:

- Conversion technologies and practices are likely to affect air quality and water quality and quantity. Examples of such impacts include findings that suggest that (1) high levels of volatile organic compounds, carbon monoxide, methanol, and other hazardous pollutants significantly affect communities with ethanol refineries; and (2) the use of dried distillers grains—a byproduct of corn-based ethanol refineries used as cattle feed—may result in microbial protein contamination, which could be harmful to human health (Figure 7).
- There are little data on the potential risks posed by leaks from storage or distribution facilities. Will the incompatibility of ethanol blends influence potential leakage from storage tanks? How do blends impact plume migration and remediation? What are the likely exposures associated with new fire retardants required to extinguish ethanol fires?
- What are the likely effects on tailpipe emissions and ambient concentrations of criteria pollutants? What are the effects of various ethanol blends on local air pollution? In particular, does it make a difference if blend levels are increased from E10 to E15 or E20?

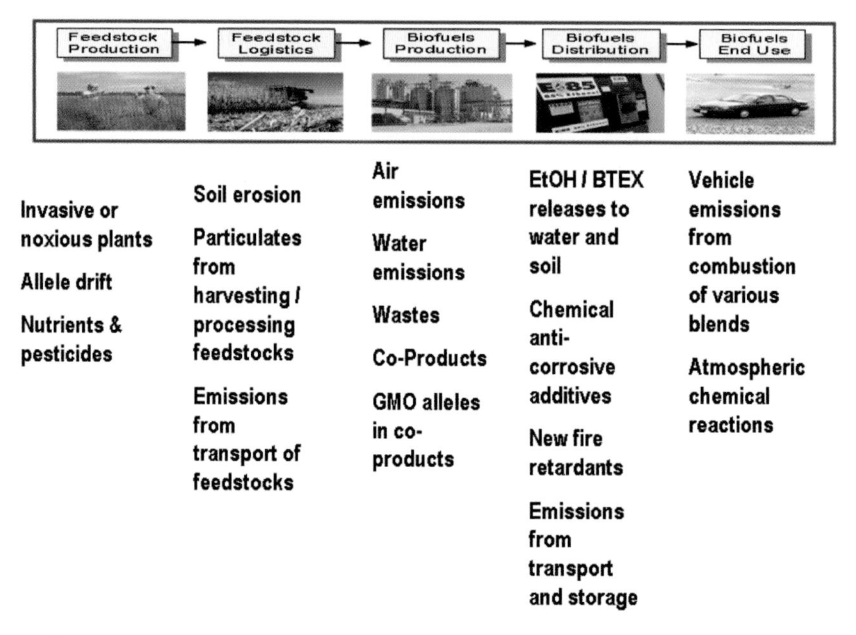

FIGURE 7 Potential releases across the supply chain, beyond GHG.
Source: Workshop Presentation by Donna Perla, U.S. Environmental Protection Agency, June 24, 2009.

One presenter proposed applying a risk framework to biofuels; identifying the environmental, health, and safety issues and benefits; integrating this information with outcomes; and comparing various potential biofuel pathways. She also advocated for more monitoring of the affected environment and of specific releases to better analyze potential risks.

SOCIAL IMPACTS

Often the social impacts associated with expanding biofuel production are not given nearly enough attention amid the gamut of highly contentious environmental and economic impacts. Current research is examining and exploring the observed and potential social impacts of the expansion of the U.S. biofuel industry. As with other industry expansions, communities and individuals will experience different impacts. The question of who may win or lose in the various scenarios was discussed by many participants—noting that rural communities that share the same values and interests are not homogeneous.

During their remarks, workshop panelists were asked to address a number of social impacts on local communities and institutions surrounding expanding biofuel production in the Upper Midwest, including:

- the impacts of the arrival or disappearance of refineries;
- the acceptability of adoption and communities' willingness to adopt new feedstocks, technologies, and fuels; and
- the impacts of changes in labor force, culture, and education.

During workshop discussions, panelists and participants raised many issues concerning the most sustainable path forward for U.S. biofuel production. While successful biofuel industry expansion in a region may be beneficial to one community (jobs, economic development, etc.), it may not be beneficial to another community with different circumstances and socioeconomic demographics. Participants also noted issues of community versus individual perceptions associated with the expansion of ethanol production, as well as unintended consequences for human health and well-being associated with negative environmental impacts.

Panel discussions highlighted the most effective ways to move forward with advanced biofuel production, while mitigating negative social impacts. Panelists and participants questioned whether the United States should repeat the same economic development policy model, or whether an alternative approach will allow for innovation coupled with a new economic development strategy for the Upper Midwest. For example, creating more holistic economic development policies at the federal and state levels that include provisions for increased energy independence and concurrently support environmental protection goals will be crucial to expanding a sustainable U.S. biofuel industry.

The issue of "winning" and "losing" was discussed extensively by participants who valued the ability to convene a much-needed, necessary and frank discussion about the kinds of tradeoffs that need to be assessed, including the impacts for winners and losers in the farming and processing communities. As the advanced biofuel industry develops, individuals—farmers who grow ethanol feedstocks and employees of refineries and processing facilities—are often perceived as winners. However, often the jobs created by ethanol production plants are not significant (e.g., fewer than 20 jobs for a smaller plant). Panelists suggested that a few new jobs may not significantly impact overall employment numbers in the Upper Midwest. Participants noted, however, that communities often believe that any new jobs are better than none.

Panelists and participants were also asked to discuss how best to minimize adverse social impacts as the industry transitions to a second generation of biofuel production. Here, many participants emphasized the need for a critical analysis of the different costs and benefits (including the path taken) in the development of the U.S. corn-based ethanol industry. Identifying the best policies and management practices will be critical to the successful development of the next-generation biofuel economy.

Many participants also emphasized the need for understanding the social and political issues of expanding a next-generation biofuel industry. How the costs and benefits will be distributed within communities was cited as an area that needs further research and attention—especially more focused data on how communities will benefit or suffer from future losses.

V

Going Forward

Much of the discussion at the workshop focused on the uncertainty and potential risks associated with expanding biofuel production and the need to look beyond a single technology solution to meet the nation's long-term needs for energy. This chapter briefly reiterates common themes that were emphasized at the workshop by many participants including: uncertainties and risks, current policies, and suggestions by participants on how future policies might be structured to ensure more sustainable energy and agricultural systems. It also describes ongoing research and existing analytical tools to address some of the current uncertainties, and includes ideas, given by participants, for additional research and tools.

As stated earlier, many participants noted that there is considerable uncertainty about future directions in the biofuel industry with regard to federal and state policies, feedstocks and technologies, financing, and energy markets. Corn-based ethanol currently accounts for 93 percent of domestic biofuels, and soybean-based biodiesel accounts for the remaining 7 percent. Because of EISA's provisions, corn-based ethanol production is not likely to grow much beyond the cap of 15 billion gallons, or 6 million gallons above current production levels. Cellulosic or other advanced biofuels are projected to account for the bulk of the expansion of biofuel production during the next 5 to 10 years, and a variety of incentives is provided to encourage development of an advanced biofuel industry. In order to increase domestic production of advanced biofuels (through the advancement of the biofuel industry from non-food crops), the Biomass Crop Assistance Program (BCAP) was created under the 2008 Farm Bill to support the production and conversion of feedstocks for bioenergy. BCAP attempts to establish greater certainty for feedstock growers and biofuel producers. The program will establish annual payments to offset risks for biomass growers and will cover most of the costs of preparing the land and planting the crops.[1] Biofuel producers can receive similar payments through BCAP to cover the costs of collection, harvest, storage, and transportation of biomass from fields to processing facilities.

Discussions on U.S. energy policy were particularly fervent during the workshop, in part, because at the time the U.S. Congress was debating major climate legislation. Most participants expressed frustration with the current lack of an integrated U.S. energy and climate policy with clear goals and objectives. They described current policies as often inconsistent, not always based on the best science, and often perversely influencing markets. While the EISA provisions requiring some biofuels to meet GHG targets were applauded, many participants were disappointed by the act's failure to create any incentives for corn-based ethanol producers to reduce emission levels or encourage performance improvements.

Similarly, EISA's failure to require that new production meet standards beyond those set for GHG emissions was seen, by a number of workshop participants, as problematic from a sustainability perspective. For example, EISA does not set targets for water efficiency. The prospect that new climate legislation would override EPA's decisions to include indirect land-use change as part of the calculation of GHG emissions was seen as a direct assault on science, since research studies have made it clear that expansion of land used for biofuel production will result in some indirect effects. While agreeing that these effects are difficult to measure, these participants pointed out that they need to be recognized.

Others suggested that more effective U.S. energy policies should be based on clear measures of performance, rather than incentivizing the production of particular energy feedstocks and technologies. Such policies would allow industry freedom and flexibility to innovate and tailor products to specific goals, such as fuel efficiency or reduced carbon emissions.

There continues to be considerable uncertainty about future feedstocks and technologies. While a number of possible feedstocks have been touted as environmentally preferable to corn and as effective sources for making advanced biofuels, participants raised many questions:

- What are the best feedstocks for particular soil, water, and climatic conditions?
- How much more difficult will it be to transport cellulosic feedstocks to refineries?
- Will these new energy crops compete for land now used for food crops?

[1]The Minnesota Project. 2009. *Transportation Biofuels in the United States: An Update*. St. Paul, MN. Available at *http://www.mnproject.org/pdf/TMP_Transportation-Biofuels-Update_Aug09.pdf*.

- Will the water and fertilizer requirements for cellulosic and other advanced biofuel feedstocks actually be significantly less than for corn?
- If these new crops are grown on "marginal" lands, will this affect the provision of valuable ecosystem services?
- Will farmers be willing to switch from traditional crops, such as corn, to new energy crops?
- Wood wastes are generally seen as an abundant and environmentally preferable feedstock for ethanol production. How will tradeoffs be made between their use for fuel or power, especially given the widespread state adoption of renewable power standards?

Several presenters talked about possible future technologies for the production of fuels from biomass—including green gasoline and other "drop-in" fuels—that can use the same distribution and storage infrastructure as petroleum based fuels. Questions were raised about what the time frame would be for the commercialization of these fuels, whether the development of these fuels would compromise activities to commercialize cellulosic fuels, and whether and how such development would affect investments in the distribution and storage systems required for ethanol or other alcohol fuels.

Participants often mentioned the inherent uncertainty in markets for biofuels, citing in particular recent decreases in demand, driven largely by relatively low prices for oil. In the longer term, the "blending wall" was seen as a constraint on demand, because the volumes of ethanol projected to be produced under EISA are far higher than can be consumed by the current fleet of flex-fueled vehicles or with gasoline mixtures of only 10 percent. Furthermore, effective demand is constrained by inadequate distribution systems with few outlets for E85, although recent allocations of federal stimulus money intend to change this to some extent. It was also noted that early investors in the ethanol industry obtained large returns over a very short period of time, while later investors were not as fortunate. In fact, many investments in new refineries failed, thus discouraging future investments both in the corn-based industry and for advanced biofuels. Another problem for investors is that while there are federal production mandates for advanced biofuels, these requirements can be waived—again creating market uncertainty.

Many participants noted that the most important environmental and social impacts associated with current corn-based ethanol and advanced biofuels are also an area of some uncertainty. Small-scale field assessments and general ecological theory suggest that cellulosic and other advanced biofuels are likely to be better from an environmental perspective than corn-based fuels, but large-scale field experience is limited. While biofuels have been touted as a boon for rural communities, the impacts on local employment and communities appear mixed.

Two workshop sessions described specific activities underway to address some of the uncertainties associated with expanded biofuel production and to develop indicators and other decision-support tools that could be used to assess at least some of the environmental impacts associated with current and future biofuel production and to evaluate tradeoffs.

Representatives from USDA, DOE, and the U.S. Geological Survey (USGS) briefly described examples of specific R&D activities currently underway.[2] They also explained the important role being played by the federal Biomass Research and Development Board to coordinate all federal agency activities and to guide future activities. The board currently has working groups focused on feedstock production, conversion science and technology, sustainability, logistics, environment, health and safety, distribution infrastructure, and blending.

USDA supports a wide variety of research related to bioenergy, including activities focused on land availability, soil suitability, climate variability, crop growth and production capacity, natural resource quality, and production practices. Research on feedstocks includes studies of region-specific varieties and practices to optimize harvest yields and on-farm use of biorefinery co-products.

The DOE representative explained that much of DOE's budget is focused on technology deployment and building pilot commercial-scale biorefineries, often partnering with industry. DOE also does basic R&D technology development. Sustainability-related activities have increased substantially, but still constitute a relatively small portion of the R&D budget. The representative noted that DOE is focusing more on advanced fuels, such as green diesel and green gasoline, which are true replacement drop-in tools for petroleum-based fuels. The DOE Regional Biomass Feedstock Partnership program is focused primarily on feedstock production. The program is a partnership with the Sun Grant universities to conduct field trials exploring ways to maximize yields and minimize inputs, whether water or nutrients, as well as reduce soil erosion.

USGS is developing sophisticated models to examine how climate change and the expansion of biofuel production in the U.S. Northern Plains will change future landscapes and the ability to provide critical ecosystem services. The study is designed to demonstrate tradeoffs and unintended consequences.

Examples of various indicators, models, and other analytical tools focused on biofuels and their sustainability implications were discussed.[3] Several different criteria and indicators efforts were reviewed, with participants questioning how they were going to be used and whether their use might result in more sustainable outcomes. It was also noted that indicators must be seen as voluntary tools to gauge performance, not as mandatory standards that might be considered nonmarket barriers to trade.

Many participants suggested that better tools and monitoring data were needed to guide policy and investment decisions as well as to measure the actual

---

[2]Detailed descriptions of specific projects sponsored by these agencies as well as a number of other federal agencies are included in Appendix C of this report.

[3]See Appendix E for descriptions of some of the major indicator efforts and models available to assess biofuels.

impacts of policy and technology choices.[4] More comprehensive systems frameworks are needed to assess the interconnected environmental, economic, and social impacts associated with particular feedstocks, conversion technologies, feedstock land choices, location of refineries, and characteristics and conditions of local environmental resources. Tools should also assess the critical linkages between energy and climate change, they said. For example—how will the productivity of land used for biofuel feedstocks be affected by climate change? In addition, participants noted the need for better tools to understand the consequences of changes in land use and to value ecosystem services that may be affected by such changes. Some participants also stressed the importance of recognizing the high degree of uncertainty inherent in current biofuel modeling efforts. To create and implement the most effective biofuels policies, decision makers should be made aware of these uncertainties.

Participants identified a number of high-priority areas for data monitoring, future research, and metrics, including the following:

- Better monitoring data, especially to assess the water quality and water quantity effects of increasing production of cellulosic and other advanced biofuel feedstocks.
- Research on the impacts of expanded production on individuals, communities, states, and regions; the extent to which local ownership increases the vulnerability of local communities; and how different ownership patterns encourage or discourage innovation and enhanced environmental stewardship.
- Research to determine whether and how extensively the biofuel industry supports rural economic development and the job creation.
- Systems analyses, linking energy and agricultural land-use change.
- Withdrawal of land from the CRP, which may affect biological diversity and ecosystem services.
- The health effects of biofuel production and use along the entire supply chain, including the effects of changes in water quality from increased use of agricultural chemicals (e.g., fertilizers and pesticides) to changes in air quality from tailpipe emissions.
- Tools to comprehensively assess sustainability impacts, including examining tradeoffs, evaluating alternative land uses, valuing ecosystem services (water, soils, biodiversity, etc.), and measuring industry performance.
- Analysis of the performance of different biofuel production technologies in fuels across all of the different parameters.

[4]These participants noted that while more comprehensive assessments are needed, those measurements are quite costly and will require significant financial investment in research.

COMMON THEMES GOING FORWARD

During the last session of the workshop breakout discussion groups were asked to discuss common themes that they saw as particularly relevant throughout the two days of panel presentations and dialogue and report back to all participants. Following are themes and issues that many participants saw as relevant and in need of further attention as the U.S. biofuel industry continues to expand.

- *Uncertainties*
  As previously stated, a common theme among workshop participants was increasing concern about the uncertainties and potential risks associated with expanding biofuel production. Choices regarding the most sustainable feedstock technology, the type of fuel to produce, and a timeline for commercial-scale production pose critical uncertainties.
- *Understanding Unintended Consequences, Value of Science*
  The majority of participants noted that much of the underlying scientific information on biofuel production is subject to variations—often times based on dynamic factors such as climate change, ecosystem health and increased efficiencies in feedstock and conversion technologies. Furthermore, examining the direct consequences of expanding biofuel production is not enough. Even more fundamentally, policy makers and scientists should be made aware of the best methods of incorporating scientific data into analytical tools and indicators for sustainability to measure and mitigate potential unintended consequences.
- *Performance-Based Metrics and Standards*
  Many participants stressed the need for clearly stated policy goals and performance-based standards, and the value of developing standards which are flexible and adaptable over time. The standards will be crucial in monitoring and evaluating global ecosystem services in the future.
- *Complex Systems and Linkages*
  Throughout the biofuel supply chain, there are many linkages between systems. To ensure sustainable biofuel production in the future, the impacts among various systems will need to be assessed. Participants noted that unintended consequences will arise when energy, water, and climate change systems are linked. Better tools and indicators can mitigate many unintended consequences of expanding biofuel production, and can enable researchers to illustrate to policy makers the effectiveness of a systems approach.
- *Full Range of Potential Impacts and Tradeoffs*
  As U.S. biofuel production expands, there will be tradeoffs that will require policy makers to make tough decisions. Most participants agreed that better tools and indicators from the research community will be crucial in assessing tradeoffs more holistically—e.g., incorporating the impacts from land and water use, biofuel production, food production,

and carbon sequestration in climate change adaptation and mitigation policies.

- *Most Critical Research Needs*
  Workshop participants focused a great deal of their discussions on future research needs as the U.S. biofuel industry expands. Such areas as human health and well-being, social and community impacts, and infrastructure needs and distribution networks were identified as requiring more in-depth research.

Appendixes

Appendix A

Workshop Agenda

Expanding Biofuel Production: Sustainability and the Transition to Advanced Biofuels

Lessons from the Upper Midwest for Sustainability

Date: June 23-24, 2009

Location: The Lussier Family Heritage Center
3101 Lake Farm Rd., Madison, WI 53711

Workshop Objectives:

- Create an opportunity for dialogue between researchers and policy makers on the sustainability impacts of expanding biofuel production at a state/regional level.
- Explore the lessons that can be learned from the experience with corn-based ethanol and the likely impacts of advanced biofuels.
- Identify biofuel-related policy objectives and challenges facing state officials.
- Provide examples of research that may be useful to state decision makers.
- Evaluate various tools and indicators that may be of use to state policy makers in assessing likely sustainability impacts and tradeoffs of policy choices.

Tuesday, June 23, 2009

LUSSIER FAMILY HERITAGE CENTER

9:00 AM **Introduction**
Emmy Simmons, Co-Chair, Roundtable on Science and Technology for Sustainability, The National Academies

Welcome
Molly Jahn, Dean, College of Agricultural and Life Sciences, University of Wisconsin

9:15 AM **Workshop Overview**
Gary Radloff, Wisconsin State Department of Agriculture

SETTING THE STAGE

9:30 AM **U.S. Energy Independence and Security Act (EISA), Implications for State Biofuels Policies**
Paul Argyropoulos and Bruce Rodan, U.S. Environmental Protection Agency

9:45 AM **Overview: Regional Biofuels Policies (Wisconsin, Minnesota, and Iowa)**
Brendan Jordan, Great Plains Institute
Judy Ziewacz, Director, Wisconsin Office of Energy Independence
John Yunker, Office of the Legislative Auditor, Minnesota

10:15 AM Discussion

10:30 AM Break

10:45 AM **A Transition to Advanced Biofuels: Where Are We?**
John Miranowski, National Research Council Panel Member, *Report: Alternative Liquid Transportation Fuels*
John Regalbuto, National Science Foundation, *Federal Inter-Agency Biomass R&D Board, Conversion Technologies Assessment Report*

11:15 AM Questions and Discussion

11:30 AM **Sustainability and a Transition to Advanced Biofuels**
John Sheehan, University of Minnesota, Institute on the Environment

- The Economy—economics of production, economic benefits, effects on other industries.
- Affected Environment—water quality and quantity, watersheds, air quality and health, soil erosion/nutrient-

level changes (including cross-media effects), land-use changes, habitat protection (including agroforestry and wood energy crops).
- Social Impacts—effects on local communities and institutions of expanding production arrival or disappearance of refineries; acceptability/willingness to adopt new fuels/technologies; changes in labor force, culture, education.

12:00 PM Questions and Discussion

12:30 PM Lunch

REGIONAL IMPACTS OF BIOFUEL PRODUCTION AND USE IN THE UPPER MIDWEST

1:30 PM **The Economics of Expanding Biofuel Production in the Upper Midwest (Panel Discussion)**
Panel Moderator: Bruce Babcock, Center for Agricultural and Rural Development, Iowa State University
What have we learned from experience with corn-based ethanol? What is required to make the industry viable going forward? What are likely impacts of expanding the production and deployment of advanced cellulosic biofuel technologies on state economies, employment, agricultural production, and prices for land and agricultural commodities? How will these impacts differ with various feedstocks? What are the likely impacts on competing users for land and biomass feedstocks (food, feed, fiber, and other bioenergy feedstocks)?
- David Swenson, Iowa State University
- Randall Fortenbery, University of Wisconsin
- Doug Tiffany, University of Minnesota

2:30 PM **Social and Community-Level Impacts of Biofuel Production in the Upper Midwest (Panel Discussion)**
Panel Moderator: Michael Bell, University of Wisconsin-Madison
What social impacts have been observed and might be seen in the future? How can adverse social impacts be minimized as we move forward with a transition to advanced biofuels? Who benefits and who stands to lose in various production scenarios, including transition from corn-based ethanol to cellulosic ethanol?
- Carmen Bain, Iowa State University
- LeAnn Tigges, University of Wisconsin

- Jim Kleinschmit, Institute for Agriculture and Trade Policy (IATP)

3:30 PM **The Environment and Health (Panel Discussion)**
Panel Moderator: Phil Robertson, GLBRC and Michigan State University
What have been the environmental impacts of expanded corn ethanol production in the Upper Midwest, and what are the likely impacts of expanding production of both cellulosic biofuels and corn-based ethanol? What does this mean for environmental sustainability in this region, and what are appropriate metrics and indicators? Discussion to include land-use changes.

- Chris Kucharik, University of Wisconsin
- Donna Perla, U.S. Environmental Protection Agency
- Peter Nowak, University of Wisconsin

4:30 PM **Breakout Discussions: Lessons Learned and a Transition Forward**
Each breakout group of participants will be asked to answer the set of questions below based on their expertise and information presented during the workshop's earlier sessions.

- Identify a comprehensive set of potential impacts associated with a transition to advanced biofuels:
- List potential environmental impacts (both positive and negative)
- List potential economic impacts (both positive and negative)
- List potential social/cultural impacts (both positive and negative)
- What are the potential strategies for mitigating potential negative consequences/negative impacts of a transition to advanced biofuels?
- What are the greatest uncertainties as we move forward with advanced biofuels (e.g., winning feedstocks)?

5:15 PM **Breakout Groups Report Back**

5:30 PM Adjourn

6:00-7:30 PM Reception, Hosted by the Wisconsin Bioenergy Initiative, Nelson Institute for Environmental Studies, and the College of Agricultural and Life Sciences, Brocach Irish Pub, 7 W. Main Street, Second floor, Madison, WI

Wednesday, June 24, 2009

LUSSIER FAMILY HERITAGE CENTER

9:00 AM **The Business of Biofuels: Perspectives from the Investment Community and Industry (Panel Discussion)**
Moderator, Pat Atkins, Pegasus Capital Advisors
- Ruth Scotti, BP Biofuels, North America
- Paul Batcheller, PrairieGold Venture Partners
- Bruce Babcock, Iowa State University

10: 00 AM Questions and Discussion

10:30 AM Break

10:45 AM **Research for Decision Makers (Panel Discussion)**
Moderator: Elisabeth Graffy, U.S. Geological Survey
What are some examples of federal research relevant to sustainability in the Upper Midwest? Additional examples of relevant research related to sustainability and biofuels in the region?
- Jeffery Steiner, U.S. Department of Agriculture
- Alison Goss Eng, U.S. Department of Energy, Biomass Program
- Alisa Gallant U.S. Geological Survey
- Theresa Selfa, Kansas State University

11:45 AM Questions and Discussion

12:00 PM Lunch

1:00 PM **Tools to Inform Policy Choices (Panel Discussion)**
Moderator: Jason Hill, University of Minnesota
What tools are available to inform policy choices? What are the strengths and weaknesses of existing tools?
- Marilyn Buford, U.S. Forest Service
- Alan Hecht, U.S. Environmental Protection Agency
- Greg Nemet, University of Wisconsin
- Nathanael Greene, NRDC

2:00 PM **Breakout Session: State Policy Objectives and Research Needs: Going Forward**
Each breakout group of participants will be asked to answer the set of questions below based on their expertise and information presented during the workshop's earlier sessions.
- Is there a need for new state/federal policies?
- What is the most pressing type of additional research needed by state decision makers?

- Are there examples of policy inconsistencies that create inefficiencies and hinder the adoption of more sustainable production techniques and use of biofuels?
- How can scientific research be better used to inform the adoption of sustainable production practices during the transition to advanced biofuels?

3:00 PM **Breakout Groups Report Back**

3:30 PM Break

3:45 PM **Breakout Session: Policy Evaluation of Tradeoffs, Benefits, and Challenges: Going Forward**
Each breakout group of participants will be asked to answer the set of questions below based on their expertise and information presented during the workshop's earlier sessions.

- What are greatest risks and vulnerabilities associated with expanded production and use?
- What will be necessary (beyond technology development) to commercialize production and use of advanced biofuels?
- What are greatest challenges (e.g., getting farmers to plant new crops, reducing risks to investors)?

4:45 PM **Breakout Groups Report Back**

5:15 PM **Wrap Up: Summary of Workshop Discussions**
Moderator: Gary Radloff, Wisconsin State Department of Agriculture

5:30 PM Adjourn

Appendix B

Workshop Participants

Pul Argyropolous
Office of Transportation and Air Quality
U.S. Environmental Protection Agency

Patrick Atkins
Pegasus Capital Advisors

Bruce Babcock
Iowa State University

Carmen Bain
Iowa State University

Paul Batcheller
PrairieGold Venture Partners

Timothy Baye
University of Wisconsin-Extension

Michael Bell
University of Wisconsin-Madison

Bill Berry
Conservation Communications

Shoshana Blank
University of Minnesota

Marilyn Buford
U.S. Forest Service

Dan Card
Minnesota Pollution Control Agency

Peter Ciborowski
Minnesota Pollution Control Agency

Alison Coulson
UW-Madison Nelson Institute

Charles Dunning
U.S. Geological Survey

Steven Fales
Iowa State University

Michael Fienen
U.S. Geological Survey

Randall Fortenberry
University of Wisconsin-Madison

Alisa Gallant
U.S. Geological Survey

Bjorn Gangeness
University of Minnesota

Alison Goss Eng
Office of Biomass Program
U.S. Department of Energy

Elisabeth Graffy
U.S. Geological Survey

Nathanael Greene
Natural Resources Defense Council

Charles Griffith
Ecology Center

Alan Hecht
Office of Research and Development
U.S. Environmental Protection Agency

Paul Heinen
Wisconsin Department of Natural Resources

Jason Hill
University of Minnesota

Tracey Holloway
University of Wisconsin-Madison

Molly Jahn
University of Wisconsin-Madison

Eric Jensen
Izaak Walton League of America

Ed Jepsen
Wisconsin Department of Natural Resources

Matt Johnston
University of Wisconsin-Madison

Brendan Jordan
Great Plains Institute

Roger Kasper
Wisconsin Department of Agriculture, Trade and Consumer Protection

Jim Kleinschmit
Institute for Agriculture and Trade Policy

Pat Koshel
National Academy of Sciences

Chris Kucharik
University of Wisconsin-Madison

Kathleen McAllister
National Academy of Sciences

Mark McDermid
Wisconsin Department of Natural Resources

Micah McMillan
U.S. General Accountability Office

Cynthia Meyer
U.S. EPA Region 5

John Miranowski
Iowa State University

Marina Moses
National Academy of Sciences

Jeff Mullins
Environmental Resources Center

Greg Nemet
University of Wisconsin-Madison

Phuong Nguyen
U.S. Environmental Protection Agency, Region 5

Peter Nowak
University of Wisconsin-Madison

Robin O'Malley
The Heinz Center

Julia Olmstead
Institute for Agriculture and Trade Policy

Andy Olsen
Environmental Law & Policy Center

Donna Perla
U.S. Environmental Protection Agency

Pamela Porter
Biomass Energy Resources Center

Gary Radloff
Wisconsin Department of Agriculture, Trade and Consumer Protection

Maria Redmnond
Wisconsin Office of Energy Independence

John Regalbuto
National Science Foundation

Phil Robertson
GLBRC and Michigan State University

Bruce Rodan
Office of Science and Technology
U.S. Environmental Protection Agency

Troy Runge
Wisconsin Bioenergy Initiative

Ruth Scotti
BP Biofuels

Theresa Selfa
Kansas State University

John Sheehan
University of Minnesota

Emmy Simmons
U.S. Agency for International Development (retired)

Jason A. Smith
The Wisconsin Academy of Sciences, Arts and Letters

Jeff Steiner
U.S. Department of Agriculture

David Swenson
Iowa State University

Peter Taglia
Clean Wisconsin

Chris Tessum
University of Minnesota

Doug Tiffany
University of Minnesota

LeAnn Tigges
University of Wisconsin-Madison

Lisa Vojta
U.S. General Accountability Office

Jeffrey Voltz
Wisconsin Department of Natural Resources

William Walker
Department of Agriculture, Trade and Consumer Protection

Sara Walling
Wisconsin Department of Agriculture, Trade and Consumer Protection

David Webb
Wisconsin Department of Natural Resources

John Yunker
Minnesota Office of the Legislative Auditor

Ronald Zalesny Jr
U.S. Forest Service, Northern Research Station

Judy Ziewacz
Wisconsin Office of Energy Independence

Appendix C

Description of Agency Activities on Biofuels and Sustainability

NOTABLE EXAMPLES OF FEDERAL RESEARCH ACTIVITIES RELATED TO BIOFUELS AND SUSTAINABILITY

TITLE OF PROJECT OR PROGRAM: Office of the Biomass Program

AGENCY: Department of Energy

PROJECT/PROGRAM DESCRIPTION:

Biomass research has been a cornerstone of DOE's renewable energy research, development and deployment efforts over the past 25 years. In order to encourage the economic livelihood of a thriving biofuel industry, the Office of the Biomass Program (OBP) at the Department of Energy supports research and development aimed at assessing the impacts of biofuels on the environment, including impacts to land, water, and air from energy production and use. Included in this mission is a goal to substantially reduce greenhouse gas emissions by accelerating the adoption of renewable energy technologies.

A clear driver of the OBP's activities is the mandate set by the Renewable Fuel Standard (RFS) which sets a U.S. production goal of 36 billion gallons of renewable fuels by 2022, of which 21 billion should be advanced biofuels made from biomass products other than corn starch, such as cellulose, algae, and waste materials. Meeting this goal will require: significant and rapid advancements in biomass feedstock and conversion technologies; availability of large volumes of sustainable biomass feedstock; demonstration and deployment of large-scale integrated biofuels production facilities; and biofuels infrastructure development efforts. In addition, the existing agricultural, forestry and commercial sectors will be making the decisions to invest in biomass systems—from shifting land use, to building capital-intensive biorefineries, to establishing the infrastructure and public vehicle fleet for ethanol distribution and end use—in the context of economic viability (including as it relates to environmental sustainability) and the needs of the marketplace.

The core R&D of OBP is organized around the integrated biorefinery concept. The biorefinery helps deliver sustainable and environmentally sound contributions to power, fuels, and products demand while supporting rural economies. Key barriers relevant to this area include ensuring resource sustainability at levels large enough to support large-scale production facilities and maximizing the efficiency of conversion facilities to minimize costs. Energy production from biomass on a large scale will require careful evaluation of U.S. agricultural resources and logistics, as these will likely require a change in paradigm that will take time to implement. Current harvesting, storage and transportation systems are currently inadequate for processing and distribution of biomass on the scale needed to support dramatically larger volumes of biofuels production. Evaluating the current feedstock resource on a national level as well as the potential for future feedstock production in light of environmental constraints is part of OBP's focus.

Overall, the program emphasizes sustainable development of the biofuels industry, including economic, environmental, and societal impacts over entire life cycle of biofuels—from the farm to end use in vehicles. The program promotes biofuels that do not compete with food crops, and our analytic models are continuously enhanced to improve our ability to anticipate, understand, and avoid potential adverse impacts on the environment, whether they are direct or indirect.

RESULTS, OUTCOMES, OR IMPACTS TO DATE:

OBP has been working with Oak Ridge, Argonne, and Idaho National Laboratories in conjunction with university partners to develop a national, GIS-based framework to analyze the economic and environmental impacts of various development options for biomass feedstocks, biorefineries, and infrastructure. The framework is aimed at supporting assessment of relevant resources and infrastructure at local, regional, and national scales; determining the best locations for new feedstock production and processing facilities; evaluating the potential contribution of biofuels to meet legislated renewable fuel production targets; and protecting air quality, water, land, and other resources.

In addition, the program's current sustainability activities include: performing comparative life-cycle assessment (LCA) of water requirements for the production of advanced biofuels, corn ethanol, sugar cane ethanol, and competing petroleum fuels. The four main areas addressed in the LCA are: land use and soil sustainability, water use impacts, air quality impacts, and greenhouse gas (GHG) emissions impacts. Also, the GREET model (*G*reenhouse gases, *R*egulated *E*missions, and *E*nergy use in *T*ransportation) is being utilized for an analysis of water demand for biofuel production, energy and GHG emission benefit of biofuels. Included in this project is an expansion of the existing model to include corn ethanol, sugarcane ethanol, and flex-fuel vehicle (FFV) test results.

Currently, LCA of the Advanced Energy Initiative is being performed for the 60 billion gallon 30x30 scenario (a scenario for supplying 30 percent of 2004 motor gasoline demands by 2030). The analysis covers the entire biofuels supply chain from feedstocks to vehicles and will expand the GREET model to incorporate other pathways including sugar cane ethanol production.

OBP is working with Conservation International to identify land that should not be developed into biofuel crops; conducting pilot studies to identify the best lands for biofuel crop production; employing standards for biofuel crop production to maintain biodiversity. The Biomass Program works with diverse partners to promote sustainable biofuels development.

OBP also participates in the Council for Sustainable Biomass Production *www.csbp.org* aimed at developing principles for bioenergy feedstocks, and as well as in the Federal Biomass Research & Development Board Interagency Sustainability working group charged with developing criteria and indicators for sustainable biofuel production.

A significant amount of work is being undertaken at Argonne National Laboratory, Oak Ridge National Laboratory, and at National Renewable Energy Laboratory to address various aspects of biofuels LCA. In addition to our ongoing support and expansion of the GREET model at Argonne, we are co-funding work on the Global Trade and Agriculture Project (GTAP) model at Purdue University. Our work at Purdue is an attempt to develop a better understanding and begin to analytically assess the indirect land use change impacts of biofuels. We continue to work with our counterparts to develop appropriate GHG accounting methodology and related policy for biofuels to enhance the climate and economic benefits of biofuels.

PERFORMERS/OTHER PARTNERS (FEDERAL, STATES, OR LOCAL):

OBP's R&D has led the effort to develop technology necessary to sustainably produce, harvest, and convert a variety of biomass feedstocks, as well as to deploy the resulting biofuels. Core R&D on feedstock production and logistics and biomass conversion technologies is conducted to develop the scientific and technical foundation that will enable the new bioindustry. OBP is looking to advance science in these areas through important collaborations with the DOE Office of Science Bioenergy Centers, the U.S. Department of Agriculture, land grant universities, and private industry. OBP has developed Regional Feedstock Partnerships to begin to realize the sustainability of the resource potential outlined in the Billion Ton Study. This approach facilitates the collaboration of industry, the agricultural community, state and local governments and USDA and is expected to accelerate the resource readiness, as the cellulosic fuels industry emerges.

PROJECT PERIOD: Ongoing

FUNDING LEVELS (CURRENT OR PROPOSED): $12.3 million in FY2008/2009; $10 million planned for FY2010

NOTABLE EXAMPLES OF FEDERAL RESEARCH ACTIVITIES RELATED TO BIOFUELS AND SUSTAINABILITY

TITLE OF PROJECT OR PROGRAM: Analysis Driven Design of Agronomic Strategies Supporting Sustainable Agricultural Residue Collection for Bioenergy

AGENCY: Department of Energy

PROJECT/PROGRAM DESCRIPTION:

The goal of this work is to build an enterprise level analysis toolset that helps design agronomic management strategies facilitating sustainable agricultural residue harvest.

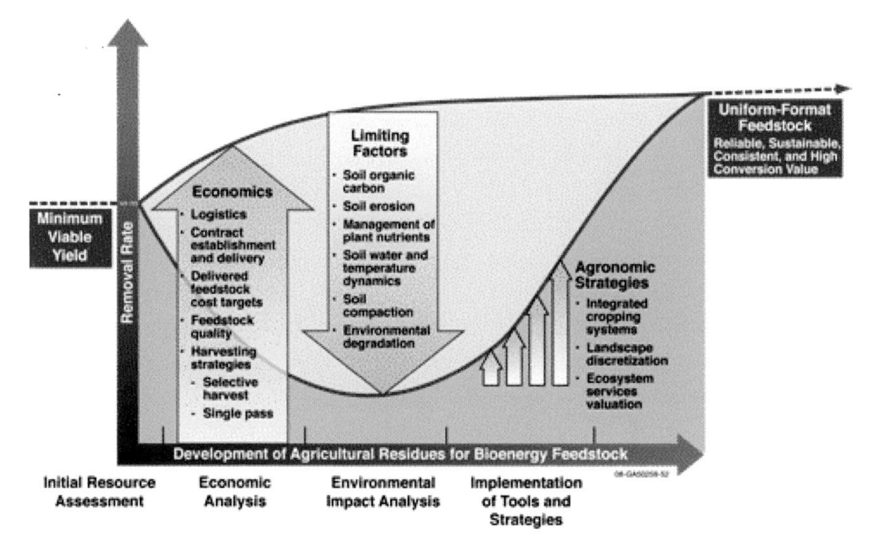

Source: U.S. Department of Energy.

Multiple factors impact agricultural residue harvest for bioenergy production. A minimum level of residue removal is required to satisfy baseline economic and logistic constraints, and increasing yield enhances viability of agricultural residues as a bioenergy feedstock. Agronomic and environmental limiting factors in many production systems reduce sustainable access to residues. The design and implementation of innovative agronomic management strategies can address sustainability issues increasing access to agricultural residues supporting biofuel production goals.

Limiting Factor Analysis Approach

Determining sustainability of residue removal within an agronomic system requires analysis taking into account the full suite of factors which limit residue removal. Each land unit has unique physical and management characteristics that determine the factor(s) impacting residue removal sustainability. The graphic above identifies the limiting factors.

Advanced Software Framework

This project is using innovative tools for software and data integration to assemble the limiting factor models in a "drag and drop" environment. Models can be pulled in and out of the system through simple interfaces facilitating analysis with the appropriate set of tools. Through this framework, individual land units can be investigated to design agronomic management strategies that provide sustainable and consistent access to residue resources.

RESULTS, OUTCOMES, OR IMPACTS TO DATE:

The figure below represents a case study demonstrating the value and importance of the analysis approach being implemented in this project. This particular run of the integrated model set is looking at a 25 acre experiment that is part of the DOE Regional Biomass Feedstock Partnership network of field trials. The site is on highly productive central Iowa soils. As demonstrated in the figure, through currently widely used analysis approaches looking at erosion alone as the limiting factor full removal of the stover residue falls within the sustainability limits for both conventional and no tillage scenarios. When the soil organic carbon limit-

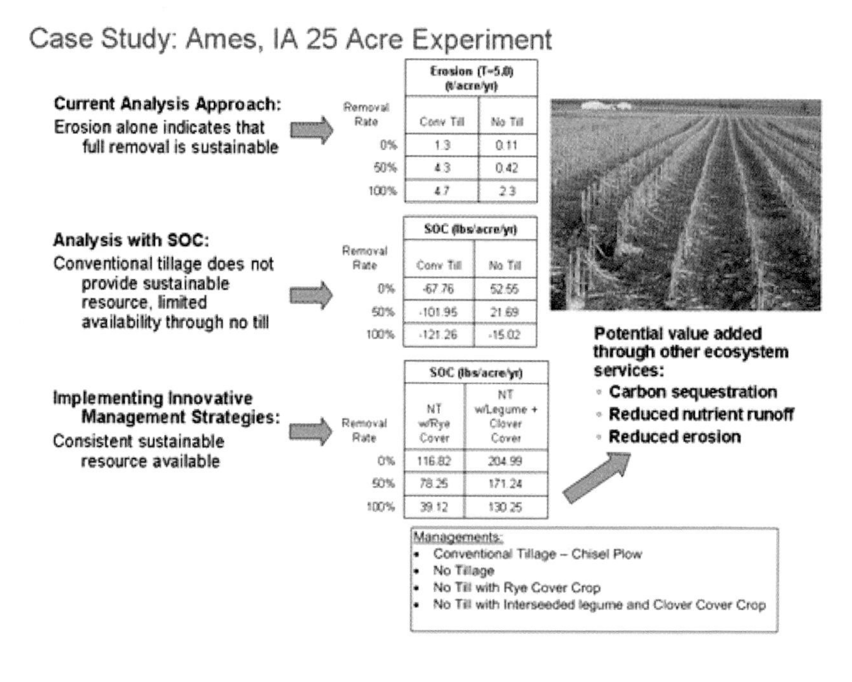

Source: U.S. Department of Energy.

ing factor is considered, no residue is sustainably accessible under conventional tillage, and partial removal is acceptable for no tillage management. Through the implementation of innovative management strategies within the analysis full residue removal is not only acceptable, but provides a soil carbon gain. This approach is working toward including each of the previously identified six limiting factors, and plans going forward include developing the ability to quantify key ecosystem services provided through the innovative strategies to potentially provide growers with added value for sustainable agronomic management.

PERFORMERS/OTHER PARTNERS (FEDERAL, STATES, OR LOCAL):

Sun Grant Initiative, Iowa State University, Idaho National Laboratory, Penn State University, Kansas State University, and USDA ARS.

PROJECT PERIOD: 1/15/07 through 9/30/10

FUNDING LEVELS (CURRENT OR PROPOSED): Current funding at 400K per year.

NOTABLE EXAMPLES OF FEDERAL RESEARCH ACTIVITIES RELATED TO BIOFUELS AND SUSTAINABILITY

TITLE OF PROJECT OR PROGRAM: Regional Biomass Feedstock Partnership Sustainability Indicator Data Collection Field Trials

AGENCY: Department of Energy

PROJECT/PROGRAM DESCRIPTION:

This project is utilizing the DOE Regional Biomass Feedstock Partnership network of field trials to begin collecting sustainability data regionally for multiple feedstock production systems. The Regional Feedstock Partnership is a multi-agency consortium comprised of land-grant universities through the Sun Grant Initiative, DOE Office of the Biomass Program, DOE National Laboratories, and USDA partners through the Agricultural Research Service and Forest Service. Among the charges of the partnership is a nationwide network of field trials assessing and developing biomass feedstock resources. This project is leveraging five of these field trials to collect data relative to critical sustainability indicators.

Eddy Covariance Tower St. Paul, MN
SOURCE: U.S. Department of Energy

Sustainability Data

Three primary sustainability indicators have been selected as critical for the specific biomass production systems being investigated are:

- Soil Carbon
  - Sequestration potential
  - Impact on productive capacity
- Hydrology and Water Quality
  - Field scale implementation
  - Nutrient transport
  - Water holding capacity
- Direct Green House Gas Emissions
  - N_2O flux
  - CO_2 flux

The Field Trials

Projects at 5 locations:

- Ames, IA; St. Paul, MN (*corn*)
- Brookings, SD (*switchgrass*)
- Champaign, IL (*miscanthus*)
- College Station, TX (*energy sorghum*)

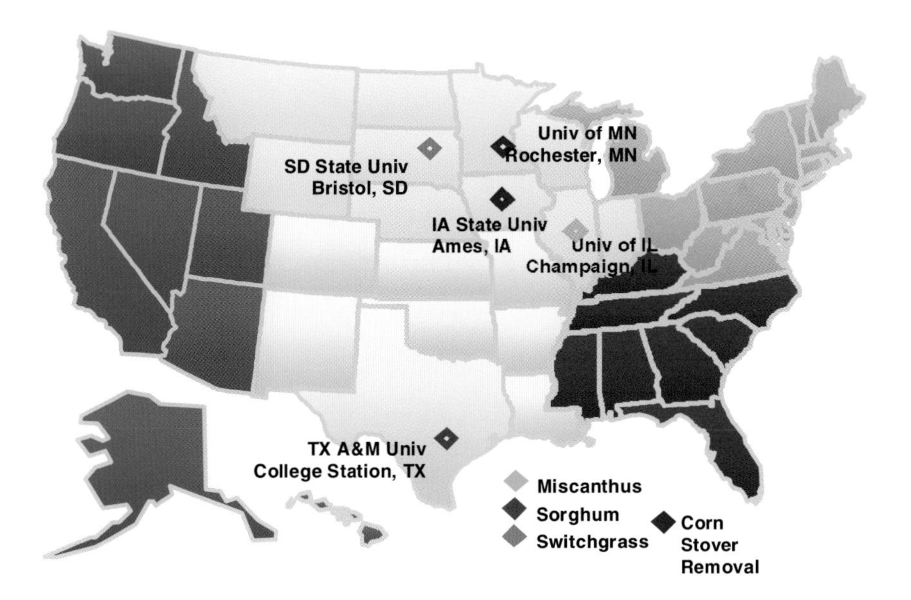

Source: U.S. Department of Energy.

The suite of feedstocks being investigated through this study will provide important data helping understand ecosystem impacts of production decisions reacting to emerging biofuel markets. Specifically of interest for the overall Regional Partnership effort, is how dedicated energy crops can integrate with currently cropping systems to provide food, feed, fiber, and fuel across an efficient and sustainable agronomic landscape. This work is focusing on developing quality sustainability based data that can inform the design of this landscape. As part of the Regional Partnership efforts, the data and publications generated through this work will disseminated through an education and outreach component of the partnership. Furthermore, the data will become part of partnership wide analyses assessing resource potential, and will be contributed to the DOE Bioenergy Knowledge Discovery Framework (KDF). The KDF is a comprehensive geospatial data and analysis toolkit being assembled to provide stakeholders with a means to interact with reviewed, up to date, and complete information about the emerging biofuels industry. The data contributed from this work will be a critical component in providing that toolkit.

RESULTS, OUTCOMES OR IMPACTS TO DATE:

The project began in January, 2009, so first year data will not be assembled until Fall, 2009. Innovative experimental designs and protocols have emerged through the planning and buildup to this project. Techniques for collecting hydrology and GHG data have been designed with associated experimental protocols for the specific implementations and will be published over the coming months and years.

PERFORMERS/OTHER PARTNERS (FEDERAL, STATES, OR LOCAL):

Sun Grant Initiative (providing a large consortium of land grant universities), USDA ARS, Idaho National Laboratory, and Oak Ridge National Laboratory

PROJECT PERIOD: 1/15/09 through 9/30/13

FUNDING LEVELS (CURRENT OR PROPOSED): Current funding at 400K per year.

NOTABLE EXAMPLES OF FEDERAL RESEARCH ACTIVITIES RELATED TO BIOFUELS AND SUSTAINABILITY

TITLE OF PROJECT OR PROGRAM: U.S. EPA's Future Midwestern Landscapes Study

AGENCY: Environment Protection Agency

Agency Contact Information:

Randy Bruins
(*bruins.randy@epa.gov*, 513-569-7581)

Betsy Smith
(*smith.betsy@epa.gov*, 919-541-0620)

Project period: 2009-2013

The Future Midwestern Landscapes (FML) Study will examine projected changes in landscapes and ecosystem services[1] in the Midwest. Given its immediate influence, biofuel production will be studied as a primary driver of landscape change. The study goals are to: (1)Understand how current and projected land uses affect the ecosystem services provided by Midwestern landscapes; (2) Provide spatially explicit information that will enable EPA to articulate sustainable approaches to environmental management and; (3) Develop web-based tools depicting alternative futures so users can evaluate trade-offs affecting ecosystem services.

For a 12-state region of the Midwest (EPA Regions 5 and 7 plus the Dakotas; Figure 1), researchers will work with decision makers and use economic and spatial modeling tools to construct alternative landscapes that reflect different assumptions about national policy, technology, and land management over the next 10-20 years.

As a first step in this project, a *Base Year* landscape has been created that represents a "pre-biofuels" scenario. To provide the level of detail necessary for relating land cover to provision of services, the National Land Cover Database (NLCD) for 2001/2002 for the region was augmented with the National Agricultural Statistical Survey (NASS) Cropland Data Layers (CDL) available for the states in the regions, soils data, and data from the LandFire database (*http://www.landfire.gov*). The new base year landscape reflects crops planted as well as typical rotations and forest species.

The *Biofuel Targets* future scenario is implied by current policies emphasizing large increases in biofuels production, as specified under the 2007 Energy Independence and Security Act (EISA). EISA calls for a ramp-up of biofuels from 2008 to 2022, beginning with increases in corn starch ethanol and later including cellulose-based ethanol, derived from a variety of sources such as corn stover, wood chips and switchgrass. Under this scenario corn production will increase,

[1]Ecosystem services can be defined as the benefits that humans derive from ecosystems.

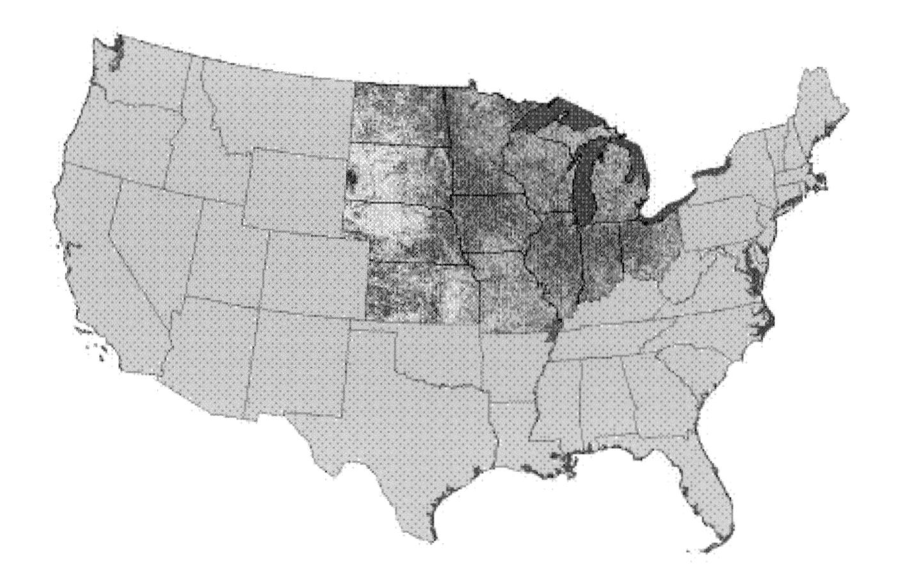

FIGURE 1 The FML Study area with the National Landuse/Landcover Database 2001/2002.
Source: U.S. Environmental Protection Agency.

primarily through conversion of other crops to corn and through modification of traditional crop rotations that alternate corn with soybeans or other crops towards a corn monoculture. There will also be a reduction in enrollment in land conservation programs, and corn stover will be the primary feedstock for cellulosic ethanol. This future landscape, which will be analyzed to evaluate the increased pressure on soil and water quality and other ecosystem services, will reflect a configuration that could be realized in 2022 under these conditions.

The alternative *Multiple Services* scenario envisions incentive policies aimed at encouraging the production of a more balanced set of agricultural commodities and environmentally beneficial outcomes. Incentives will tend to favor enhanced agronomic and conservation practices that provide societal benefits such as water quality, flood control, carbon storage and wildlife production. The suite of ecosystem services that are provided by Midwestern landscapes will be individually weighted by representative decision-makers using a multi-criteria decision analysis method to develop an optimal target landscape for the region. Next, hypothetical incentive policies will be crafted that would support landowners' adoption of those conservation practices that best support the broader suite of services. An economic model will then assess the efficacy of the proposed policies and the results will be used to adjust the optimal landscape towards a more realistic, spatially-explicit representation at the year 2022.

The ecosystem services associated with each alternative landscape will be described and compared. Ecosystem services we will seek to assess include:

- Carbon sequestration (affects climate)
- Soil productivity (affects food and energy security)
- Hydrology and water quality (affect water supply, flooding, downstream aquatic ecosystems, recreation)
- Wildlife habitat and other natural areas (affect biodiversity and recreation)
- Air quality (affects health)

Evaluating many ecosystem services will require that we draw upon the expertise of other federal agencies. Collaboration is underway with the USDA Farm Service Agency, USDOI Fish and Wildlife Service, and the U.S. Army Corps of Engineers.

The landscape analysis methods developed for the study will be implemented as a web-based environmental decision toolkit, similar to other toolkits previously created under EPA's Regional Vulnerability Assessment Program (ReVA). Scientists anticipate that the toolkit will allow users to compare alternative Midwestern futures by examining tradeoffs—that is, changes in the provision of a wide variety of ecosystem services—at both local and regional scales.

NOTABLE EXAMPLES OF FEDERAL RESEARCH ACTIVITIES RELATED TO BIOFUELS AND SUSTAINABILITY

TITLE OF PROJECT OR PROGRAM: Lifecycle Analysis of Greenhouse Gas Emissions from Renewable Fuels

AGENCY: U.S. Environmental Protection Agency

PROJECT/PROGRAM DESCRIPTION:

As part of proposed revisions to the National Renewable Fuel Standard program (commonly known as the RFS program), EPA analyzed lifecycle greenhouse gas (GHG) emissions from increased renewable fuels use. The Energy Independence and Security Act of 2007 (EISA) establishes new renewable fuel categories and eligibility requirements. EISA sets the first U.S. mandatory lifecycle GHG reduction thresholds for renewable fuel categories, as compared to those of average petroleum fuels used in 2005. The regulatory purpose of the lifecycle GHG emissions analysis is to determine whether renewable fuels meet the GHG thresholds for the different categories of renewable fuel.

Lifecycle GHG emissions are the aggregate quantity of GHGs related to the full fuel cycle, including all stages of fuel and feedstock production and distribution, from feedstock generation and extraction through distribution and delivery and use of the finished fuel. The lifecycle GHG emissions of the renewable fuel are compared to the lifecycle GHG emissions for gasoline or diesel (whichever is being replaced by the renewable fuel) sold or distributed as transportation fuel in 2005.

EISA established specific GHG emission thresholds for each of four types of renewable fuels, requiring a percentage improvement compared to a baseline of the gasoline and diesel. EISA required a 20 percent reduction in lifecycle GHG emissions for any renewable fuel produced at new facilities (those constructed after enactment), a 50 percent reduction in order to be classified as biomass-based diesel or advanced biofuel, and a 60 percent reduction in order to be classified as cellulosic biofuel. EISA provides some limited flexibility for EPA to adjust these GHG percentage thresholds downward by up to 10 percent under certain circumstances. EPA is proposing to exercise this flexibility for the advanced biofuels category in this proposal.

EPA must conduct a lifecycle analysis to determine whether or not renewable fuels produced under varying conditions will meet the GHG thresholds for the different fuel types for which EISA establishes mandates. While these thresholds do not constitute a control on GHGs for transportation fuels (such as a low carbon fuel standard), they do require that the volume mandates be met through the use of renewable fuels that meet certain lifecycle GHG reduction thresholds when compared to the baseline lifecycle emissions of petroleum fuel. Determining compliance with the thresholds requires a comprehensive evaluation of renewable fuels, as well as of gasoline and diesel, on the basis of their lifecycle emissions. EISA defines lifecycle GHG emissions as follows:

> The term 'lifecycle greenhouse gas emissions' means the aggregate quantity of greenhouse gas emissions (including direct emissions and significant indirect emissions such as significant emissions from land use changes), as determined by the Administrator, related to the full fuel lifecycle, including all stages of fuel and feedstock production and distribution, from feedstock generation or extraction through the distribution and delivery and use of the finished fuel to the ultimate consumer, where the mass values for all greenhouse gases are adjusted to account for their relative global warming potential.[2]

As mandated by EISA, the GHG emission assessments must evaluate the full lifecycle emission impacts of fuel production including both direct and indirect emissions such as significant emissions from land use changes. We recognize the significance of using lifecycle GHG emission assessments that include indirect impacts such as emission impacts of indirect land use changes. Therefore, in our proposal we have been transparent in breaking out the various sources of GHG emissions to enable the reader to readily detect the impact of including international land use impacts.

RESULTS, OUTCOMES, OR IMPACTS TO DATE:

EPA has analyzed the lifecycle GHG impacts of the range of biofuels currently expected to contribute significantly to meeting the volume mandates of EISA through 2022, including those from domestic and international sources. In these analyses we have used the best science available. Our analysis relies on peer reviewed models and the best estimate of important trends in agricultural practices and fuel production technologies as these may impact our prediction of individual biofuel GHG performance through 2022. We have identified and highlighted assumptions and model inputs that particularly influence our assessment and seek comment on these assumptions, the models we have used and our overall methodology so as to assure the most robust assessment of lifecycle GHG performance for the final rule.

The GHG lifecycle analysis combines a suite of peer-reviewed process models and peer-reviewed economic models of the domestic and international agricultural sectors to determine direct and significant indirect emissions, respectively (see Figure 1). As required by EISA, the broad system boundaries of our analysis encompass all significant secondary agricultural sector GHG impacts, not only impacts from land use change. The analysis uses economic models to determine the area and location of land converted into cropland in each country as a result of the RFS program. Satellite data are used to predict the types of land that would be converted into cropland (e.g., forest, grassland).

EPA's draft results suggest that biofuel-induced land use change can produce

---

[2]Clean Air Act Section 211(o)(1).

significant near-term GHG emissions; however, displacement of petroleum by biofuels over subsequent years can "pay back" earlier land conversion impacts. Therefore, the time horizon over which emissions are analyzed and the application of a discount rate to value near-term versus longer-term emissions are critical factors. We highlight two options. One option assumes a 30 year time period for assessing future GHG emissions impacts and values equally all emission impacts, regardless of time of emission impact (i.e., 0 percent discount rate). The second option assesses emissions impacts over a 100 year time period and discounts future emissions at 2 percent annually. Several other variations of time period and discount rate are also discussed in the proposed rule. Table 1 provides draft GHG emission reductions that result under two time horizon/discount rate approaches for a sample of fuel pathways evaluated in the proposed rulemaking. Figures 1 and 2 break out emissions for each of these pathways by lifecycle component (e.g., fuel production, domestic and international and use change, domestic and international agricultural inputs) for the two time horizon/discount rate approaches.

We believe that our lifecycle analysis is based on the best available science, and recognize that in some aspects it represents a cutting edge approach to addressing lifecycle GHG emissions. Because of the varying degrees of uncertainty in the different aspects of our analysis, we conducted a number of sensitivity analyses which focus on key parameters and demonstrate how our assessments might change under alternative assumptions. By focusing attention on these key parameters, the comments we receive as well as additional investigation and

TABLE 1 Draft Lifecycle GHG Emission Reduction Results for Different Time Horizon and Discount Rate Approaches

Fuel Pathway	100 year, 2% Discount Rate	30 year, 0% Discount Rate
Corn Ethanol (Natural Gas Dry Mill)	–16%	+5%
Corn Ethanol (Best Case Natural Gas Dry Mill)[a]	–39%	–18%
Corn Ethanol (Coal Dry Mill)	+13%	+34%
Corn Ethanol (Biomass Dry Mill)	–39%	–18%
Corn Ethanol (Biomass Dry Mill with Combined Heat and Power)	–47%	–26%
Soy-Based Biodiesel	–22%	+4%
Waste Grease Biodiesel	–80%	–80%
Sugarcane Ethanol	–44%	–26%
Switchgrass Ethanol	–128%	–124%
Corn Stover Ethanol	–115%	–116%

[a]Best case plants produce wet distillers grain co-product and include the following technologies: combined heat and power (CHP), fractionation, membrane separation and raw starch hydrolysis.
Source: U.S. Environmental Protection Agency.

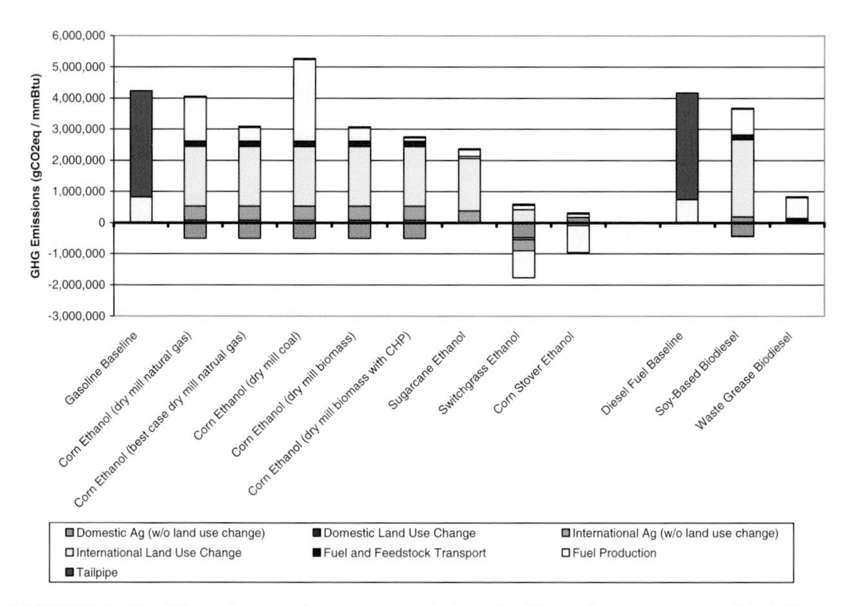

FIGURE 1 Net lifecycle greenhouse gas emissions by lifecycle component with 100 year time horizon and 2% discount rate.
Source: U.S. Environmental Protection Agency.

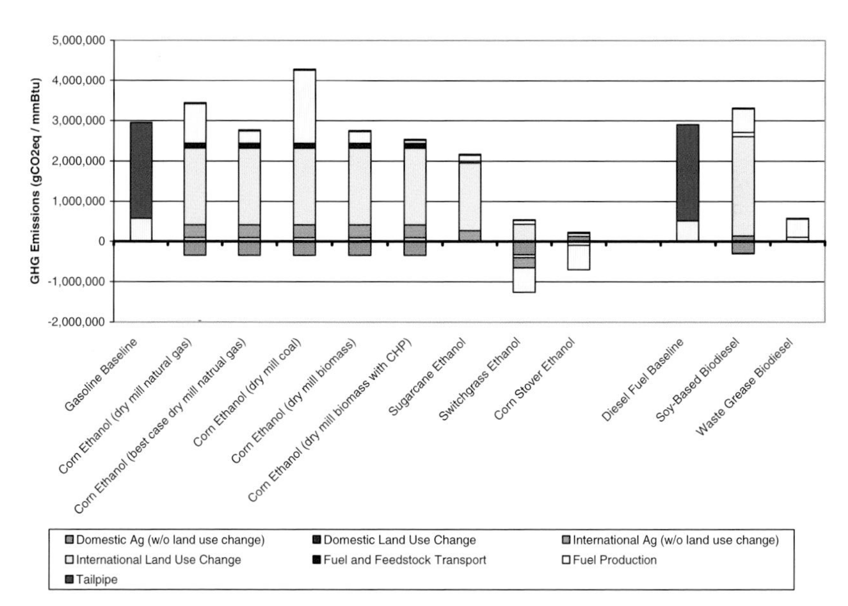

FIGURE 2 Net lifecycle greenhouse gas emissions by lifecycle component with 30 year time horizon and 0% discount rate.
Source: U.S. Environmental Protection Agency.

analysis by EPA will allow narrowing of uncertainty concerns for the final rule. In addition to this sensitivity analysis approach, we will also explore options for more formal uncertainty analyses for the final rule to the extent possible.

Because lifecycle analysis is a new part of the RFS program, in addition to the formal comment period on the proposed rule, EPA is making multiple efforts to solicit public and expert feedback on our proposed approach. EPA plans to hold a public workshop focused specifically on lifecycle analysis during the comment period to assure full understanding of the analyses conducted, the issues addressed and the options that are discussed. We expect that this workshop will help ensure that we receive submission of the most thoughtful and useful comments to this proposal and that the best methodology and assumptions are used for calculating GHG emissions impacts of fuels for the final rule. Additionally, between this proposal and the final rule, we will conduct peer-reviews of key components of our analysis. As explained in more detail in the section VI of the proposal, EPA is specifically seeking peer review of: our use of satellite data to project future the type of land use changes; the land conversion GHG emissions factors estimates we have used for different types of land use; our estimates of GHG emissions from foreign crop production; methods to account for the variable timing of GHG emissions; and how the several models we have relied upon are used together to provide overall lifecycle GHG estimates.

Each component of our analysis is discussed in detail in the preamble and the Draft Regulatory Impact Analysis that accompany the Notice of Proposed Rulemaking. The proposed rule is an important opportunity to seek public comment on EPA's entire lifecycle GHG analysis, including questions about land use modeling, and the choice of which time horizon and discount rate is most appropriate for this analysis.

PERFORMERS/OTHER PARTNERS (FEDERAL, STATES, OR LOCAL): Information not available

PROJECT PERIOD: Information not available

FUNDING LEVELS (CURRENT OR PROPOSED):
Information not available.

FOR MORE INFORMATION, PLEASE VISIT:
www.epa.gov/otaq/renewablefuels/index.htm

NOTABLE EXAMPLES OF FEDERAL RESEARCH ACTIVITIES RELATED TO BIOFUELS AND SUSTAINABILITY

TITLE OF PROJECT OR PROGRAM: Multi-scale Assessment of Environmental and Economic Sustainability for Renewable Biomass Production Systems

AGENCY: USDA-Agricultural Research Service

PROJECT/PROGRAM DESCRIPTION:

The project utilizes multi-location field research to refine process-based models Environmental Policy Integrated Climate (EPIC) and Agricultural Land Management Alternatives with Numerical Assessment Criteria (ALMANAC), allowing field-validated simulations of potential biofuel crop species. Field-level simulation is used to determine on-farm sustainability including production, profitability and economic risk, break-even biomass prices, nitrate and pesticide leaching, sediment and nutrient run-off, wildlife value, and soil carbon storage. Results are spatially-linked, allowing display of results through a geographic information system, providing capability to map results and conduct spatial analyses of biomass supplies and resulting economic and environmental impacts.

Field data are scaled to the watershed level with the Soil Water Assessment Tool (SWAT). Analyses address offsite sustainability issues including water supply, river and reservoir sedimentation, and nutrient and pesticide loading and concentrations in rivers and reservoirs. The system will also ultimately assess the impact of biofuel feedstock production on hypoxia in the Gulf of Mexico. The approach emphasizes field-level detail that cannot be simulated using large-scale empirical models. Results at the local, regional, and national scales will be calculated with high accuracy by aggregating and routing simulations based on a field-parameterized, field-validated lowest unit.

Regional-scale economic analyses utilize Parallel Genetic Algorithm for Computation of Biophysical and Economic many-objective Pareto Sets (PGA-BIOECON), an analysis approach that dynamically links economic and physical simulation models. This provides an efficient mechanism for calculating optimal tradeoffs among objectives including farm profitability, policy efficiency, and ecological services. The system uses data envelopment analysis (DEA) to model producer behavior at the farm-level. PGA-BIOECON requires actual farm level data that has been obtained with a novel approach that uses Bayesian methods to generate synthetic microdata from Census of Agriculture individual records. This method preserves the statistical characteristics of the original farm survey data and allows economic modeling and mapping of results, yet preserves confidentiality of census records. The use of DEA allows the evaluation of non-market goods or "bads" (e.g., sediment or nitrogen run-off). Cap-and-trade marketing, permitting, and other incentive policies have also been modeled using DEA. To characterize environmental quality or water quality, a Malmquist index approach is implemented in the model. In this approach, the weight that is attached to each metric used in assessing environmental quality is derived from the entire data set. This approach takes advantage of over a century of results in index number theory, and meets all the theoretical requirements for an index to accurately represent the underlying metrics. PGA-BIOECON uses SWAT as the primary model for simulation of the physical environment.

RESULTS, OUTCOMES, OR IMPACTS TO DATE:

The hybrid genetic algorithm that is the basis for PGA-BIOECON was developed as part of the USDA Conservation Effects Assessment Project (CEAP). The publication of the algorithm is the first application in economics of methods that calculate a spread of points along the whole Pareto optimal front in multiple dimensions. Previously, only a single point at a time was estimated, and it was very difficult to include the interactions of variables in the search for multiple optima.

In the CEAP project, PGA-BIOECON is being applied to 12 watersheds to evaluate the trade-offs among farm profit, water quality, and program efficiency where conservation practices are adopted. This research is the first scientific attempt at assessing the environmental benefits that result from the public investment in agricultural conservation programs through the USDA Farm Bill Conservation Title.

Field-level simulations have been conducted for the entire Minnesota River watershed, and the results have been used to evaluate potential field-scale economic and environmental impacts for alternative crop production practices, and to generate regional biomass supply curves, and relate biomass supply to aggregate environmental impacts for a biomass gasification facility constructed at Morris, MN.

SWAT has been applied to the Upper Mississippi River Basin (UMRB) for estimation of climate change induced stream flow, for calculation of sediment, nitrogen and phosphorus loads due to ethanol production under different scenarios in the UMRB. SWAT is also used as the primary model for the National Assessment Project, which provides an annual accounting of the environmental benefits obtained from USDA conservation program expenditures.

It is the intent of this project to link to the USDA Economic Research Service Regional Environment and Agriculture and Programming (REAP) model so that the effects of alternative bioenergy production systems on commodity prices can be evaluated, and these results used to help optimize income at the producer level.

PERFORMERS/OTHER PARTNERS (FEDERAL, STATES, OR LOCAL):

USDA-ARS Mandan, ND, Temple, TX, and Corvallis, OR, Texas A&M University, University of Minnesota-Morris, USDA-Natural Resources Conservation

Service, Oregon State University, USDA-Economic Research Service, USDA-National Agricultural Statistics Service.

PROJECT PERIOD: Long-term research, specific accomplishment timelines: 2007-2010.

FUNDING LEVELS (CURRENT OR PROPOSED): Recurrent USDA-Agricultural Research Service base funding.

NOTABLE EXAMPLES OF FEDERAL RESEARCH ACTIVITIES RELATED TO BIOFUELS AND SUSTAINABILITY

TITLE OF PROJECT: Managing Agricultural Drainage Systems for Economic and Environmental Sustainability of Biofuels Production

AGENCY: USDA-Agricultural Research Service

PROJECT/PROGRAM DESCRIPTION:

The unique soil and climate of the Upper Mississippi River Basin (and the Lake Erie Basin) area provide the resources for bountiful agricultural production. Agricultural drainage—both surface and subsurface—is essential for achieving economically viable crop production. As agricultural producers strive to meet the demands of producing grain and biomass feedstocks for food, feed, renewable energy generation, more production will be required from each unit of land. This will likely cause the land currently not in production to be brought back into use with the consequent application of more agrichemical use in watersheds. The result will increase the potential for soluble pollutant delivery from agricultural production areas to surface and ground waters. Drainage practices alter hydrology, shortening the travel distance and time for water to move from the landscape into the stream networks, as well as increasing the volume of water moving to the streams. Consequently the water interacts less with the mineral and organic components of the soil, so there is less opportunity for biological and chemical interactions to process dissolved nutrients carried with the drainage water. Historically these were managed as free drainage systems, allowing all the water that reached the drain to flow freely to the receiving stream.

USDA Agricultural Research Service (ARS) research in Ohio documented significant reductions in drainage volume and nitrate load delivered offsite with alternative winter season drainage water management by raising the drainage outlet elevation. The advent of this concept of management led to the formation of a multi-agency USDA effort called the Agricultural Drainage Management Systems (ADMS) Task Force under the joint leadership of ARS, Natural Resources Conservation Service (NRCS), and the Cooperative State Research, Education, and Extension Service (CSREES). Task Force meetings have brought together many state and federal agencies to examine and develop ways to promote adoption of this practice in the Upper Midwest as the most promising practice for reducing the contribution from this region towards the size of the hypoxic zone in the Gulf of Mexico. This had led to adoption of drainage water management as a cost-shareable practice available to land owners and operators under the NRCS Environmental Quality Incentives Program (EQIP). Establishment of the ADMS Task Force also spawned the formation of a drainage industry-based Agricultural Drainage Management Coalition (ADMC), which has contributed important insights to the discussions and strategies developed by the Task Force. Understanding the need for demonstration of this innovative technology and education of the agricultural community to enhance its adoption, the ADMC worked with ARS and University research scientists to draft a Conservation Innovation Grant (CIG) proposal to establish demonstration sites in five Midwest states. NRCS provided funding for this CIG project, and the water quality and economic benefits of the DWM practice are being quantified, tested, and demonstrated at the field scale at more than 20 locations across the Midwest.

The primary objectives of the CIG project are to demonstrate reductions in flow and nutrient load to receiving streams, and to assess the potential yield benefit of crop season drainage water management through additional soil water available for crop use. Educational programs are being offered throughout the Midwest to promote the design and management of these innovative drainage systems, and the cost sharing programs that are available in some states as incentives for installation and management.

Field sites were selected on privately owned and managed cropland with existing subsurface drainage systems with two or more outlets. Each site drains at least 15 acres and was selected based on uniformity of soils, drainage design, and cropping management. Both outlets are equipped with a drainage water management control structure where flow and nitrate concentration are monitored. One outlet is maintained in a free drainage mode and the other outlet is elevated during the non-growing season to within 1 foot of the soil surface. All management inputs are uniform over the entire field allowing quantification of the hydrologic and water quality effects of the drainage water management practice. There are currently more than 20 monitored sites located on different soil types and cropping management systems across the Upper Midwest region in Illinois, Indiana, Iowa, Minnesota, and Ohio. Geographic Positioning System referenced combine yield monitors record the spatial yield effects, and the growers provide complete input records so that an economic evaluation of the drainage water management practice can be made.

RESULTS, OUTCOMES OR IMPACTS TO DATE:

- Depending on local soil, climate and management conditions, annual subsurface drainage flows have been reduced 30 to 65 percent where drainage water management was applied.
- Drainage water management practice standards have been revised for all Midwest states and approved for cost-share payments under EQIP.
- A $1,000,000 CIG has been received to demonstrate and evaluate drainage water management in five states.
- Education programs and materials have been developed and delivered to designers, installers, operators, and agency representatives.

PERFORMERS/OTHER PARTNERS (FEDERAL, STATES, OR LOCAL):

USDA-NRCS Columbus, OH, USDA-Natural Resources Conservation Service, USDA-Cooperative States Research, Education, and Extension Service, USDA Farm Service Agency, USDA National Agricultural Statistics Service, U.S. Environmental Protection Agency, state-level Environmental Protection Agencies, The Nature Conservancy, Sand County Foundation, University of Minnesota, University of Illinois, Iowa State University, Ohio State University, Purdue University, North Carolina State University, Agricultural Drainage Management Coalition and private land owners and operators.

PROJECT PERIOD: USDA-ARS Columbus, OH research was in 1999, ADMS Task Force organized in 2003, ADMC organized in 2005, and USDA-NRCS-CIG grant received in 2006.

FUNDING LEVELS (CURRENT OR PROPOSED):

Recurrent USDA-Agricultural Research Service base funding. USDA-NRCS provided $500,000 in appropriated funds for the CIG, with matching funds came from various sources including salary and equipment donations. Now that the infrastructure exists as a result of the CIG, additional research funding on the order of $2,000,000 per year for five years is needed to collect, analyze, and interpret the data including model development and testing.

NOTABLE EXAMPLES OF FEDERAL RESEARCH ACTIVITIES RELATED TO BIOFUELS AND SUSTAINABILITY

TITLE OF PROJECT OR PROGRAM: Integrated Management Systems for Biofuel Production in the Western Corn Belt

AGENCY: USDA-Agricultural Research Service

PROJECT/PROGRAM DESCRIPTION:

Integrated systems research by USDA-ARS (ARS) at the National Soil Tilth Laboratory in Ames, IA supports the development of a future sustainable biofuels industry by developing technologies and new interpretations to quantify agricultural system impacts on soil, air and water resources for agroecosystem in the upper Midwestern region. This region is dominated by corn and soybean production for livestock and ethanol production now, and which will be expected to support future biofuels production from cellulose. Embedded within these agricultural landscapes are remnant woodlands, grasslands, and water bodies that also provide significant ecosystem services for wildlife habitat and recreation.

The ARS Integrated Management Systems for Biofuel Production in the Western Corn Belt effort contributes to a USDA nationwide effort known as the Conservation Effects Assessment Project (CEAP). The overall goal of CEAP is to quantify the impact of agricultural conservation practices on water quality. ARS and university partners supported by the USDA Cooperative States Research, Education, and Extension Service (CSREES) are conducting research and providing the technology needed for USDA Natural Resources Conservation Service (NRCS) to assess the value of conservation practice supported by USDA Farm Bill Conservation Title programs. The ARS contributions are measuring and modeling environmental impacts of agricultural and conservation practices based on research at two ARS long-term watersheds in the South Fork of the Iowa River and the Walnut Creek watersheds in central Iowa that represent typical agroecosystems for the region, but which differ in current levels of ethanol and animal production. These watersheds provide a wide array of landscape-soil-cropping system combinations for the region that allow the effects of climatic variation and management changes to be assessed relative to temporal variations. These watersheds are part of the ARS nation-wide long-term research watershed network.

On-going measurements have been made since 2002 for nutrient (N and P), sediment, and pathogen loads at eight nested locations within watershed subbasins. Water and carbon dioxide fluxes will be measured for corn and soybean fields in the spring of 2009. These data will be used to calibrate a modified Environmental Policy Integrated Climate (EPIC) crop yield model to simulate the impacts of farming practices on water balance and crop harvest index to estimate the amount of soil by an integration of the ARS SQSTR carbon sequestration model with EPIC. The models will be spatially extended using combinations of remotely sensed data acquired from RapidEye satellite and aircraft-based hyperspectral sensors to quantify the spatial variation of cropping systems across the watersheds and regions. This approach will assess the spatial variation within fields that is induced by different soil types, and provide a real-time verification of actual biomass amounts by remote sensing. Estimates of the greenhouse gas fluxes (CO_2 and N_2O) from different cropping systems and soil management practices are made with ancillary experiments that are conducted on similar soils near Ames. This information is applied to the different types of farming practices that are observed for fifty fields within the South Fork watershed.

The State of Iowa integrated watershed approach provides a framework to assess the impacts of changing soil management practices (e.g., removal of crop residue, changing nitrogen management, and changing crop rotation sequences) across a large-scale that can be challenged with multiple scenarios for weather and soils. Collection of extensive and intensive information across a watershed supplies data that are placed into a GIS-based SQL database for use in several different analyses. The Parallel Genetic Algorithm for Computation of Biophysical and Economic many-objective Pareto Sets (PGA-BIOECON) approach will contribute to the integration of components into an assessment of sustainability and ecosystem services.

Funding to conduct landscape-scale experiments on biofuel crop production under management control is needed to exploit fully the potential of this monitoring and modeling effort. Several additional field-scale sub-watersheds have been identified that could provide a sequence of experiments aimed at assessment of crop management for biofuel production and its impacts on multiple soil, water, and air quality endpoints.

RESULTS, OUTCOMES OR IMPACTS TO DATE:

Research accomplishments have focused on assessments of watershed-scale processes and effectiveness of conservation practices as distributed under current policy incentive structures.

- An assessment of nitrate, phosphorus, and bacterial contaminants has shown each contaminant to be uniquely timed, highlighting the complexity of watershed assessments. These assessments are pointing the way towards contaminant-specific conservation targeting strategies in tile-drained watersheds.
- In the Iowa River's South Fork watershed, significant water quality challenges remain despite an 80% rate of conservation-practice adoption. The key reasons for this are: (1) legacy impacts of past agricultural practices; (2) specific gaps in time of conservation effectiveness under the corn-soybean rotation; and, most importantly for nitrate loads; and (3) current conservation systems do not address the tile drainage pathway, which delivered about 70 percent of the stream discharge during a four year assessment period.
- Linkage with a local watershed group is in place and has provided multiple technology transfer opportunities. At the regional scale, linkage with

Heartland Region Water Quality Initiative has facilitated transfer of research results to extension educators and state agency personnel across EPA Region 7.

PERFORMERS/OTHER PARTNERS (FEDERAL, STATES, OR LOCAL):

USDA-ARS, Ames, IA and Beltsville, MD. Collaborations are established with local and state USDA-NRCS offices, and multiple academic departments at Iowa State University. Initial discussions are underway with U.S. Geological Survey-Upper Midwest Environmental Sciences Center for aquatic and terrestrial assessments. We have various collaborations with state agency and NGO stakeholder organizations, both agricultural and environmental. One of these projects is a collaboration to develop Revised Universal Soil Loss Equation (RUSLE2) planning tool using LIDAR-based topographic data, which should become available for the South Fork watershed within the next year. We have established a strong partnership with the South Fork Watershed Alliance.

PROJECT PERIOD: On-going long-term research-specific USDA-ARS accomplishment timelines: 2007-2012.

FUNDING LEVELS (CURRENT OR PROPOSED): Recurrent USDA-ARS base funding.

NOTABLE EXAMPLES OF FEDERAL RESEARCH ACTIVITIES RELATED TO BIOFUELS AND SUSTAINABILITY

TITLE OF PROJECT OR PROGRAM: Renewable Energy Assessment Project (REAP)

AGENCY: USDA-Agricultural Research Service

PROJECT/PROGRAM DESCRIPTION:

Recently crop residues, specifically corn (*Zea mays* L.) stover, have been identified as a primary feedstock for second-generation lingo-cellulosic biofuel production. However, success for cellulosic-based biofuels production will depend on science-based guidance that guides the sustainable harvest crop residues so that croplands will still be protected from erosion, and enhance soil organic carbon (SOC) so that the projected increases in crop productivity needed to meet market needs can be achieved. The Billion Ton Biomass Report and other publications have considered the potential water and wind erosion effects of stover harvest. However, research over the past century has shown conclusively that prevailing crop production practices often result in loss of SOC, even without stover removal. Loss of SOC has negative effects on crop productivity because of reduced soil quality.

The Renewable Energy Assessment Project (REAP) research objectives are: (1) Determine the impacts of residue removal on soil quality constituents for different cropping systems used across the United States; (2) Develop algorithms for estimating the amount of crop residue that can be sustainably harvested for different ecological regions; (3) Provide guidelines for developing management practices supporting sustainable harvest of residues; and (4) Contribute to the development of decision support tools by USDA-NRCS and others describing the economic trade-off between residue harvest and the unintended consequences of residue harvest. Delivery of these products to farmers and the emerging biomass conversion industry will promote sustainable corn stover and other crop residue harvest in a manner that preserves the capacity of our soil resource to produce food, feed, fiber, and fuel now, and in the future.

In 2008, the USDA-ARS REAP efforts were enhanced through the U.S. Department of Energy-Sun Grant Regional Partnership with researchers at Cornell University, University of Tennessee, South Dakota State University, Oklahoma State University, and Oregon State University and administered by the Cooperative States Research Extension and Education Service (CSREES). The partnership's objective are complimentary with the USDA-ARS national REAP efforts.

Grain and stover yield, changes in soil quality indicators, strategies for stover harvest (e.g., single- or multiple-pass operations), and feedstock quality and energy values are being determined. Greenhouse gas emissions and nutrient leaching measurements are being gathered where possible. Additional research sites are being added, and the USDA-ARS network of long-term research sites are being modified as needed to broaden the application of the research. Individual farmers as well as corporations including Monsanto Inc., John Deere, Inc., and POET are working cooperatively with REAP and Sun Grant Regional Partnership team members. An interdisciplinary approach is used that includes soil and plant scientists, engineers, economists, and rural sociologists from the various participating institutions are leveraged by the core activities to address broader aspects of sustainable feedstock production. The REAP effort is also supported by the USDA-ARS nation-wide research watersheds and greenhouse gases and carbon flux networks.

RESULTS, OUTCOMES OR IMPACTS TO DATE:

- The ARS process-based carbon sequestration model (CQESTR) has been adapted to predict the impact of removing residue at different rates. This model is being adapted for use with remote sensing technology and interface with the Soil Water Assessment Tool (SWAT) watershed model.
- The ARS REAP team has developed preliminary algorithms to estimate the minimal biomass inputs needed to maintain SOC for long-term soil sustainability at selected sites in the upper Midwest States region.
- Using the REAP approach, four single-pass corn stover harvest scenarios have been investigated to find the optimal harvest strategy for residue harvest. A single-pass harvesting system is being developed to gather corn grain and stover simultaneously in one harvest pass. USDA-ARS contributes soil quality assessments, Iowa State University equipment engineering, and U.S. Department of Energy (DOE) provides stover residue quality measurements.
- Core information being developed for the REAP database is being prepared for entry into the (DOE) Knowledge Development Framework databases.
- Results from the USDA-ARS REAP and Sun Grant Regional Partnership have been presented through a wide variety of media, with more than 30 entries made to the Agricultural Research Information System (ARIS) since 2005.

PERFORMERS/OTHER PARTNERS (FEDERAL, STATES, OR LOCAL):

USDA-ARS at Akron, CO, Ames, IA, Auburn, AL, Beltsville, MD, Booneville, OR, Brookings, SD, Corvallis, OR, Dawson, GA, Florence, SC, Fort Collins, CO, Lincoln, NE, Mandan, ND, Morris, MN, Orono, ME, Pendleton, OR, Prosser, WA, Pullman, WA, Saint Paul, MN, Sidney, MT, Stoneville, MS, University Park, PA, Watkinsville, GA, and West Lafayette, IN partners with university colleagues at 24 locations across the U.S., Monsanto Inc., Idaho and Oakridge National Laboratories, John Deere, Inc., and many other local agricultural, bioenergy industries including the Chippewa Valley Ethanol Cooperative.

PROJECT PERIOD: On-going long-term research; specific USDA-ARS accomplishment timelines have been developed for 2007-2012.

FUNDING LEVELS (CURRENT OR PROPOSED): Recurrent USDA-Agricultural Research Service base funding. Partial additional funding is provided by the Sun Grant Initiative to support ARS partners.

NOTABLE EXAMPLES OF FEDERAL RESEARCH ACTIVITIES RELATED TO BIOFUELS AND SUSTAINABILITY

TITLE OF PROJECT OR PROGRAM: Improved Bioenergy Plants and Production Technologies for the Central USA

AGENCY: USDA-Agricultural Research Service

PROJECT/PROGRAM DESCRIPTION:

The long-term objectives of this project are the development of improved perennial grasses and management practices to optimize the abundant and dependable supplies of biomass to biorefineries. The focus of the research is on switchgrass and other warm- and cool-season grasses that are adapted to grazinglands in the U.S. central states ecoregions. The specific objectives being addressed are: (1) Provide improved plant materials, and (2) Develop management practices and sustainable systems that maintain quality stands over multiple years of harvest; optimize biomass and net energy yield; optimize economic return for producers; and provide beneficial environmental services such as erosion control and carbon sequestration.

RESULTS, OUTCOMES OR IMPACTS TO DATE:

This location has conducted switchgrass research since 1935, with bioenergy research becoming an emphasis in 1990. This long-term research program has developed improved genetic materials in concert with the production systems needed to realize their yield potential. The unit's breeding and genetics thrust has resulted in the release of two improved switchgrass cultivars, two improved big bluestem cultivars, and three improved indiangrass cultivars, all with biomass energy potential. Switchgrass cultivars specifically developed for bioenergy will be released in late 2009.

We have developed agronomic practices and management information for the production and utilization of improved switchgrass used as a biomass energy crop, including seeding rates and seedbed preparation, herbicide tolerance, seed quality and seed dormancy, nitrogen fertility rates, harvest management, and mycorrhizal requirements.

We have developed baseline environmental and economic performance information for switchgrass grown for biomass energy. A large-scale study conducted on 10 farms in three states and five production years demonstrated that switchgrass biomass could be produced for bioethanol production with an average farm cost of $60/ton which would result in a farm gate cost of $0.64 per gallon of ethanol.

We demonstrated the effects of switchgrass composition differences due to maturity at harvest and genetic interactions on potential bioenergy conversion for both biomass to ethanol conversion and thermochemical conversion to biogases or bio-crude.

We have determined the fundamental genetic underpinnings for switchgrass to advance improvement of this species including verification of principal ploidy levels; cytogenetic behavior as diploids—enabling diploid quantitative genetic models to be used; demonstrated a gametophic self-incompatibility system; demonstrated that lowland and upland tetraploid ecotypes are completely cross-compatible; demonstrated significant high parental source heterosis for biomass yield; designed a breeding system for producing F_1 hybrids; and developed a publically available, complete EST genomic profile in collaboration with ARS scientists at Albany, CA.

We applied the Plant Adaptation Region concept to classify plant germplasm by ecogeographic regions, and validated the concept as a mechanism for defining adaptation regions for switchgrass cultivars based on origin of ecological types.

Our study on net energy sustainability demonstrated that switchgrass could produce 13 times more energy in the form of ethanol than would be required as energy from petroleum, and produced 540% more renewable energy than non-renewable energy consumed on marginal land, when properly managed.

We have demonstrated that within five years of production, significant amounts of carbon are sequestered in soils growing switchgrass, with accrual rates of 1.1 and 2.9 Mg C ha^{-1} yr^{-1}, in the 0-30 cm and 0-120 cm, respectively. With this level of soil carbon sequestration, switchgrass production can have an environmentally positive greenhouse gas profile.

We have also improved upon the switchgrass cultivars Trailblazer and Shawnee, and have new bioenergy-specific cultivars to be released in 2009. Other native grass cultivars with biomass energy potential include our Bonanza and Goldmine big bluestem, and Chief, Scout, and Warrior indiangrass.

PERFORMERS/OTHER PARTNERS (FEDERAL, STATES, OR LOCAL):

USDA-ARS Lincoln, NE. Production systems and genetic performance research is done in collaboration with USDA-ARS Temple, TX and Mandan, ND, and University of Nebraska, Iowa State University, and University of Illinois. Genetic improvement research is conducted by ARS in cooperation with the ARS Western Research Center in Albany, CA, and local partnerships with private seed growers.

PROJECT PERIOD: Ongoing long-term research, with specific accomplishment timelines: 2008-2013.

FUNDING LEVELS (CURRENT OR PROPOSED): Recurrent USDA-Agricultural Research Service base funding, with $1,000,000 per year, with half of the funding appropriated to bioenergy and half to forage and pasture.

NOTABLE EXAMPLES OF FEDERAL RESEARCH ACTIVITIES RELATED TO BIOFUELS AND SUSTAINABILITY

TITLE OF PROJECT OR PROGRAM: Land-use, Soil Health, and Water Quality Changes with Woody Energy Crop Production in Wisconsin and Minnesota

AGENCY: U.S. Forest Service, Northern Research Station, Institute for Applied Ecosystem Studies (NRS-13), Rhinelander, WI

PROJECT/PROGRAM DESCRIPTION:

Short rotation woody crops (SRWC) such as *Populus* species and hybrids (hereafter referred to as poplars) are renewable energy feedstocks that can potentially be used to offset electricity generation and natural gas use in many temperature regions, such as Wisconsin and Minnesota. Highly productive poplars grown primarily on marginal agricultural sites are an important component of our future Midwest energy strategy. Additionally, poplars can be strategically placed in the landscape to conserve soil and water, recycle nutrients, and sequester carbon. These purpose-grown trees are vital to reducing our dependence on nonrenewable and foreign sources of energy used for heat and power. Establishing poplar genotypes that are adapted to local environmental conditions substantially increases establishment success and productivity. But, it is difficult to predict field trial success in landscapes where the crop has not been previously deployed. Our overall goal is to merge our knowledge of poplar biology with large-scale spatial analysis to predefine zones of potential plant adaptation that are ecologically sustainable and economically feasible across the landscape.

The project builds on SRWC research conducted at the IAES in Rhinelander since 1968, as well as decades of poplar genetics research in Minnesota that has led to commercial poplar production on >10,000 ha in the state. Along with empirical data on poplar growth and productivity collected in both states, we will first combine key climatic and soil properties with land ownership and use constraints to develop a GIS-based spatial analysis protocol to identify candidate core areas for potential establishment. We will then construct a comprehensive poplar database and apply that information within the candidate core areas. Our final task is to evaluate land-use, soil health, and water quality changes within these areas to synthesize the environmental and social constraints on woody energy crop development within the region.

Our approach is novel in that it integrates genetics and landscape ecology, so that sustainable crop development can be more rapid, precise, and efficient. This type of approach has never been conducted for woody energy crop production. Landowners and industrial representatives will use the results of the study to evaluate trade-offs of woody energy crop production versus other uses, while researchers will benefit from the development of the protocol and availability of the soil and water synthesis that is currently not available in this region.

RESULTS, OUTCOMES OR IMPACTS TO DATE:

Decades of information relating to short rotation woody crop production; however, there are no results from the project described above as it was initiated in March 2009.

PERFORMERS/OTHER PARTNERS (FEDERAL, STATES, OR LOCAL):

Research Team

Ronald S. Zalesny Jr.,*[,1] Deahn M. Donner,[1] David R. Coyle,[2] Brian R. Sturtevant,[1] Eric J. Gustafson,[1] Neil D. Nelson,[1] and Don E. Riemenschneider[1]

[1]U.S. Forest Service, Northern Research Station, Institute for Applied Ecosystem Studies (IAES)

[2]University of Wisconsin, Department of Entomology

Collaborators (Regional Experts)

Richard B. Hall (Iowa State University, Department of Natural Resource Ecology and Management, Ames, IA)

Bill Berguson (University of Minnesota–Natural Resources Research Institute, Duluth, MN)

Raymond O. Miller (Michigan State University, Michigan Agricultural Experiment Station, Escanaba, MI)

Mark D. Coleman (University of Idaho, Department of Forest Resources, Moscow, ID)

Brian J. Stanton (GreenWood Resources, Inc., Portland, OR)

PROJECT PERIOD: March 2009-September 2011 (current funding); October 2011-? (ongoing with additional funds)

FUNDING LEVELS (CURRENT OR PROPOSED):

$169,020:	Wisconsin Focus on Energy Environmental and Economic Research and Development Program
$ 3,000:	University of Wisconsin–Madison
$102,000:	U.S. Forest Service NRS-13
$274,020:	Total

NOTABLE EXAMPLES OF FEDERAL RESEARCH ACTIVITIES RELATED TO BIOFUELS AND SUSTAINABILITY

TITLE OF PROJECT OR PROGRAM: Impacts of Harvesting Forest Residues for Bioenergy on Nutrient Cycling and Community Assemblages in Northern Hardwood Forests

AGENCY: U.S. Forest Service, Northern Research Station, Institute for Applied Ecosystem Studies (NRS-13), Rhinelander, WI

PROJECT/PROGRAM DESCRIPTION:

The most readily available source of woody biomass to the logger is through whole-tree harvesting that removes what has been traditionally left as slash (i.e., fine woody debris-FWD). This material has potential to be used as energy feedstock. However, a critical element of managing for biodiversity is maintaining woody debris on the forest floor. Woody biomass is important for nutrient cycling, providing seed beds, and creating habitat structure for wildlife. Researchers recognize the link between biodiversity and ecosystem functioning, but this relationship is not well understood. A change in species may have cascading effects across trophic levels, and cause shifts in the size, distribution, and vertical zonation of vegetation over large areas. Our goal is to investigate the impact of FWD removal on nutrient availability and above and belowground community assemblages on rich soils under regenerating northern hardwood stands in Wisconsin.

Land managers are concerned with removing FWD in this system because of the existing lack of large woody debris and structural diversity (e.g., understory shrubs). We will manipulate the amount of fine woody debris removed after timber harvest (e.g., 0, 65 and 100 percent) at 9 sites within the Chequamegon-Nicolet National Forest to compare soil carbon-nitrogen availability and other important soil physical and chemical characteristics, as well as community change (i.e., the abundance and diversity of plant, arthropod, and vertebrate assemblages) across treatments. We will test several hypotheses including: (1) soil carbon and nitrogen will decrease on sites with less woody residue, thus lowering carbon sequestration rates and nitrogen availability for regeneration, and increasing soil acidity, which could influence plant and insect communities, (2) non-native and early successional plants will increase on sites with less woody residue due to site disturbance associated with harvesting techniques, (3) seed dispersing arthropod abundance (primarily ants) will decline, while the abundance of species associated with early successional plants (e.g., invasive root-feeding weevils) will increase on sites with less woody residue, influencing overall plant diversity and forest health, and (4) frog and salamander numbers will decline on sites with less woody residue due to microclimate temperature and moisture changes, and a change in insect community assemblages.

Investigating several trophic levels simultaneously during an experimental study will help determine the underlying mechanisms behind the change in diversity effects. Study results can be used by policy makers to evaluate the trade-offs of harvesting woody biomass on pubic lands for energy against other values, and propose a set of management guidelines that can provide energy feedstocks while maintaining biodiversity and forest health.

RESULTS, OUTCOMES OR IMPACTS TO DATE:

There are no results from the project described above as we are currently establishing plots and harvesting will be conducted during winter 2009-2010.

PERFORMERS/OTHER PARTNERS (FEDERAL, STATES, OR LOCAL):

Research Team

Deahn M. Donner,[1] Matthew St. Pierre,[2] Ronald S. Zalesny Jr.,[1] Christine A. Ribic,[3] David R. Coyle,[4] and Dan Eklund[2]

[1]Institute for Applied Ecosystem Studies, Northern Research Station, Rhinelander, WI
[2]Chequamegon-Nicolet National Forest, Rhinelander, WI
[3]U. S. Gelogical Survey, WI Cooperative Wildlife Research Unit, University of WI–Madison, WI
[4]Department of Entomology, University of WI–Madison, Madison, WI

PROJECT PERIOD: July 2008-September 2011 (current funding); October 2011-? (ongoing with additional funds)

FUNDING LEVELS (CURRENT OR PROPOSED):

$144,155:	Wisconsin Focus on Energy Environmental and Economic Research and Development Program
$ 3,500:	University of Wisconsin–Madison
$302,000:	U.S. Forest Service, Chequamegon–Nicolet National Forest
$ 80,000:	U.S. Forest Service, Institute for Applied Ecosystem Studies (NRS-13)
$529,655:	Total

NOTABLE EXAMPLES OF FEDERAL RESEARCH ACTIVITIES RELATED TO BIOFUELS AND SUSTAINABILITY

TITLE OF PROJECT OR PROGRAM: Impact of Rapid Land-Use Change in the Northern Great Plains: Integrated Modeling of Land-Use Patterns, Biophysical Responses, Sustainability, and Economic and Environmental Consequences (aka "Biomass for Energy and Ecosystem Services")

AGENCY: U.S. Geological Survey in Partnership with University of Minnesota

PROJECT/PROGRAM DESCRIPTION:

We are evaluating the effects of an expanded agricultural base for biofuels and concurrent changes in climate on ecosystem sustainability across the Northern Great Plains (Figure 1). This research tests whether *land use patterns driven largely by economic considerations will be sustainable*. We are projecting alternative landscape futures at annual time steps through 2050, analyzing the results to estimate effects on ecosystem processes and services. Socioeconomic drivers, such as national policy and programs, commodity prices, and biofuel demand, are being incorporated to develop multiple scenarios that variously emphasize production of corn, soybeans, switchgrass, and mixed prairie species. We also are addressing management practices (e.g., tillage and residual biomass in soils) that we expect to have appreciable impacts on soil organic carbon, soil erosion,

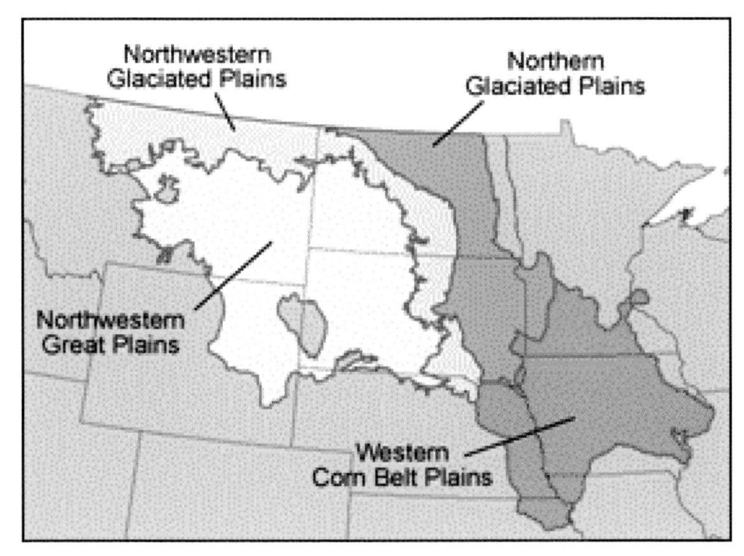

FIGURE 1 The four ecoregions of the study area.
Source: U.S. Geological Survey.

and, subsequently, water quality. Each scenario is being implemented for current climate conditions, low-change conditions, and high-change conditions, as defined by the Intergovernmental Panel on Climate Change. We are using the model *FORE*casting *SCE*narios of Land Cover Change (FORE-SCE) to develop annual maps of landscape change; the *G*eneral *E*nsemble biogeochemical *M*odeling *S*ystem (GEMS) to model biogeochemical response to land cover and land use; the *B*etter *A*ssessment *S*cience *I*ntegrating Point & *N*onpoint *S*ources (BASINS) model to estimate associated levels of soil erosion and nutrient, pollutant (e.g., nitrate), and sediment loadings to major waterbodies; economic and econometric models to determine agricultural profitability and energy costs and benefits, and a synoptic landscape analysis to estimate quantity and quality of habitat for wildlife (amphibians, birds, pollinators). We will assess environmental quality and sustainability based on total carbon accounting, agricultural productivity, greenhouse gas emissions, sediment and nutrient loadings to waterbodies, and availability and quality of wildlife habitat.

Analyses and results are being conducted at multiple scales to provide information relevant for decisions at national, regional/state, and local levels. A web-enabled decision support tool, EcoServ, has been prototyped to estimate effects of land-management and climate changes on multiple and simultaneous ecosystem services (Figure 2). The tool integrates information from a set of submodels and can query live databases across the internet, such as climate data from the National Center for Atmospheric Research. Currently, the EcoServ prototype performs local-scale analyses of responses in water storage, floristic quality, and amphibian and waterfowl habitat to changes in climate and land management. In development are submodels to provide estimates for additional ecosystem services related to soil erosion and sedimentation reduction, water quality (nutrient status), ground-water recharge, vegetation biomass, carbon sequestration, greenhouse gases, shorebirds, and pollinators.

RESULTS, OUTCOMES, OR IMPACTS TO DATE:

We are completing the first year of this multiyear project and do not yet have results to share beyond the prototype for the EcoServ model (Figure 2).

PERFORMERS/OTHER PARTNERS (FEDERAL, STATES, OR LOCAL):

U.S. Geological Survey, University of Minnesota, National Aeronautics and Space Administration (NASA), U.S. Department of Agriculture Natural Resources Conservation Service (NRCS; national level), USDA Farm Service Agency (FSA; national, state, and county levels), Chinese Academy of Science, University of Maryland, and University of Nebraska.

Principle Contributions:

U.S. Geological Survey—Landscape projections, biogeochemical modeling and carbon accounting, greenhouse gas estimation, vegetation biomass

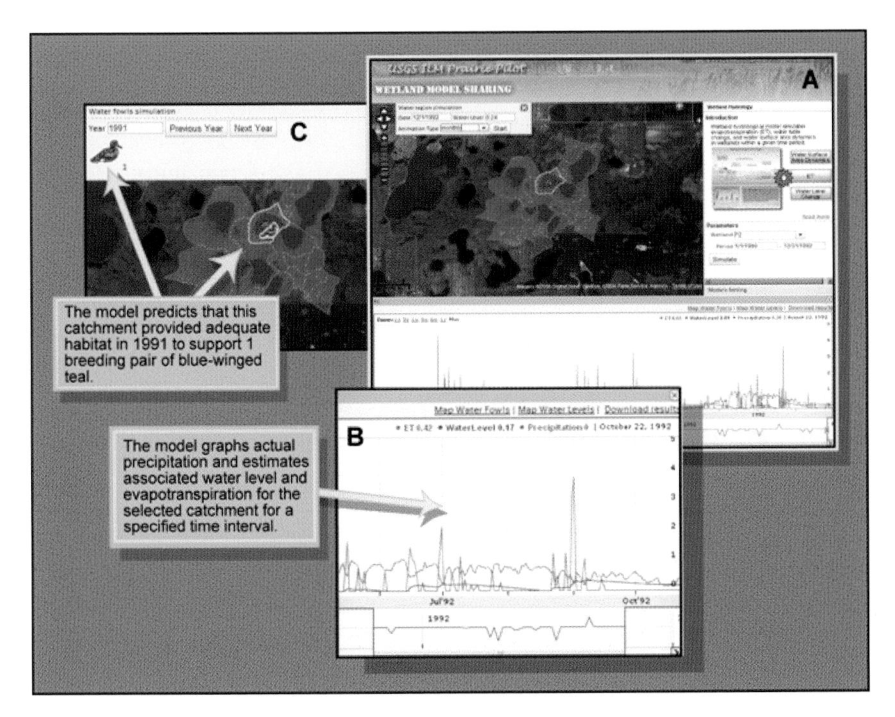

FIGURE 2 The EcoServ model currently sits on the Google Earth background (A) and accesses a set of submodels focused on specific ecosystem processes and services. In the examples shown here, the model has dynamically queried the climate archive at the National Center for Atmospheric Research (NCAR) in Boulder, Colorado, to access precipitation and temperature data for a period specified by the user. EcoServ then used a submodel to run hydrologic simulations and created graphs showing actual precipitation plotted against estimated evapotranspiration and water level for a user-selected catchment (B). Water-level information was used in conjunction with data on wetland type within a wetland model developed for the Prairie Pothole Region to estimate suitable habitat for number of breeding pairs of different waterfowl species (C).
Source: U.S. Geological Survey.

estimation, water quality, animal habitat estimation, ecosystem service valuation, and development of the EcoServ model.

University of Minnesota—Economic and econometric modeling, scenario development for landscape projections, ecosystem services valuation.

NASA—Funding support.

USDA NRCS—Funding support.

USDA FSA—Funding support and access to agricultural data previously unavailable for research.

Chinese Academy of Science—Web-enabled modeling for EcoServ

University of Maryland—Web-enabled modeling for EcoServ

University of Nebraska—Evapotranspiration modeling, ecosystem primary production, drought modeling, biogeochemical modeling.

PROJECT PERIOD: 2008-2012

FUNDING LEVELS (CURRENT OR PROPOSED): Funding across years is variable, incomplete, and highly leveraged. The funding level for FY2009 is approximately $685,000 (before USGS assessment of 45 percent), although $150,000 of this remains uncertain.

NOTABLE EXAMPLES OF FEDERAL RESEARCH ACTIVITIES RELATED TO BIOFUELS AND SUSTAINABILITY

TITLE OF PROJECT OR PROGRAM: Effects of Biofuel Development on Watershed Scale Hydrologic Flows: Scenario Testing

AGENCY: U.S. Geological Survey

PROJECT/PROGRAM DESCRIPTION:

Background: Climate change, decreases in traditional energy sources, and land- and water-use changes due to population increase will affect natural systems. Biofuel development and related changes in conservation practices such as Conservation Reserve Program (CRP) set-asides could greatly alter the current agricultural/environment balance, especially in areas that may be stressed by future expansion of urbanization. The relative importance of these stressors is not well understood, which can hamper decision-making.

Objectives: To quantify the effects of potential biofuel development on watershed scale hydrologic flows using an existing coupled ground-water/surface-water model, and relate the degree of system change due to biofuel production to that resulting from potential land use and climate change (funded by others). The effects would be evaluated in terms of single stressors, and in combination.

Approach: The proposed work would take advantage of a new constructed fully coupled ground-water/surface-water model constructed for the Black Earth Creek (BEC) watershed previously funded by USGS and non-USGS funds. A coupled model is critical for characterizing the ranges of potential stress because the feedbacks between the unsaturated zone, surface-water, and ground-water, systems are explicitly included. Thus, future scenarios can be evaluated using the effects on both storm and base flows—entities important for understanding flooding, sediment transport, and environmental low-flows, as well as related ecosystem effects such as stream temperature.

Fully coupled models are not widely available, and the BEC model is notable for being one of the first coupled models developed using the new USGS code GSFLOW (Markstrom and others, 2008). The BEC watershed is well suited to assess the effects of biofuel development because it was developed to address issues common in Midwest biofuel regions (e.g., effects of Best Management Practices (BMP), cold-water fishery and flooding issues), encompasses topography endemic to the Midwest (glaciated and non-glaciated areas), and is located in an area with a long history of study and field data collection (Figure 1).

The current model is being run to assess the effects of potential climate change, as well as future mitigated and unmitigated urbanization. The work proposed here would extend the existing model scenarios to include: (1) conversion of various landscape categories (hillslopes, CRP, current BMP) to active agriculture typical of both corn and biennial crops such as switchgrass; (2)

FIGURE 1 Black Earth Creek watershed location.
Source: U.S. Geological Survey.

increase of pumping for irrigation due to biofuel production, with and without expected climate change; and (3) high-capacity ground-water-withdrawal typical of a biofuel production plant. Landscape categories would be delineated using a combination of remote sensing, aerial photograph, and land-records analysis (Figure 2). Effects of CRP conversion would be assessed using the infiltration rates of Steuer and Hunt (2001) for an adjacent basin. Irrigation and biofuel plant water-withdrawal volumes simulated would encompass a range of reasonable literature values for the Midwest. A representative subset of all scenarios would be included in a USGS Scientific-Investigations Report being finalized in FY2010.

RESULTS, OUTCOMES, OR IMPACTS TO DATE:

We are still in the process of calibrating the GSFLOW model to properly simulate connections between groundwater and streams in relation to land use changes. The initial scenario to be tested is siting of a biofuels production plant near Black Earth Creek.

PERFORMERS/OTHER PARTNERS (FEDERAL, STATES, OR LOCAL):

U.S. Geological Survey—Wisconsin Water Science Center (WI WSC)
U.S. Geological Survey—Earth Resources Observation and Science (EROS) Data Center

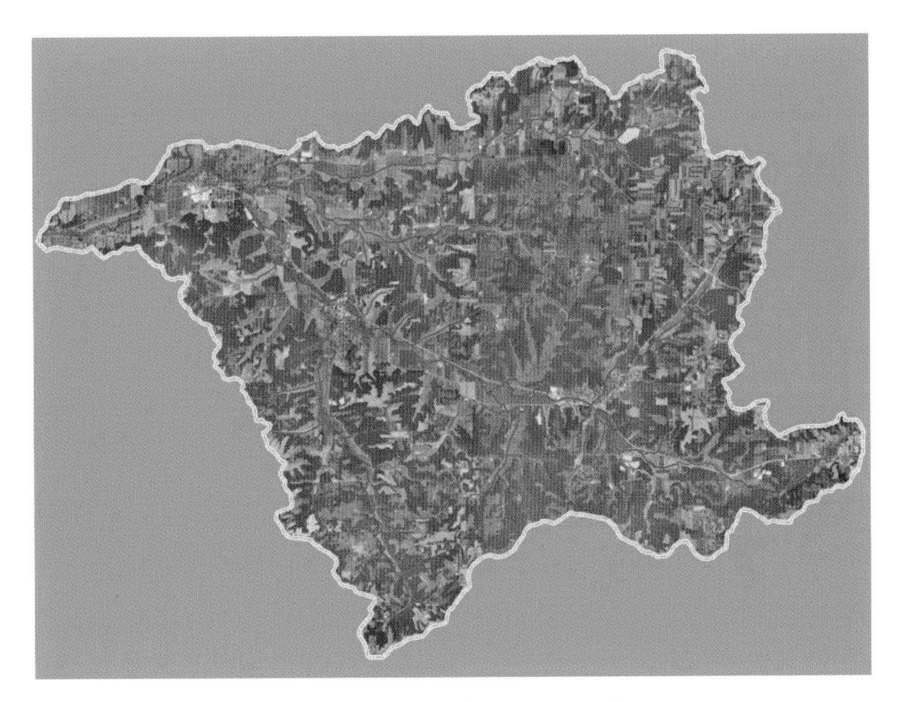

FIGURE 2 Black Earth Creek land cover and stream network.

PROJECT PERIOD: 2008-2010

FUNDING LEVELS (CURRENT OR PROPOSED): Landscape categories delineation, $15,000 (WI WSC and EROS)

Biofuel scenario model input/run/post-process, $35,000 (WI WSC)

References Cited

Markstrom, S.L., R.G. Niswonger, R.S. Regan, D.E. Prudic and P.M. Barlow, 2007, GSFLOW—Coupled Ground-water and surface water flow model based on the integration of the Precipitation-Runoff Modeling System (PRMS) and the Modular Ground-Water Flow model (MODFLOW-2005). U.S. Geological Survey Techniques and Methods 6-D1.

Steuer, J.J., and R.J. Hunt, 2001, Use of a Watershed-Modeling Approach to Assess Hydrologic Effects of Urbanization, North Fork Pheasant Branch Basin near Middleton, Wisconsin: U.S. Geological Survey Water-Resources Investigations Report 01-4113, 49 p.

NOTABLE EXAMPLES OF FEDERAL RESEARCH ACTIVITIES RELATED TO BIOFUELS AND SUSTAINABILITY

TITLE OF PROJECT OR PROGRAM: Estimation of Nutrient and Sediment Loading in the Mississippi River and Great Lakes Basins with Regional SPARROW Models

AGENCY: U.S. Geological Survey

PROJECT/PROGRAM DESCRIPTION:

There has been increasing expectations to develop and implement effective nutrient reduction strategies in the Mississippi/Atchafalaya River Basin (MARB) to reduce the size of the hypoxia zone in the Gulf of Mexico and in the Great Lakes Basin to limit the productivity in each of the Great Lakes. With support from the Environmental Protection Agency (EPA) and the National Water-Quality Assessment (NAWQA) Program, SPARROW (a hybrid statistical /mechanistic watershed model) models are being developed to explain spatial patterns in monitored stream-water quality (nutrient yields) in relation to human activities and natural processes that influence the transport of nutrients as defined by detailed geospatial information.

Results from SPARROW water-quality models are being used to describe where on the landscape nutrients originate, what are the sources of those nutrients, how watersheds rank throughout large basins in terms of their nutrient

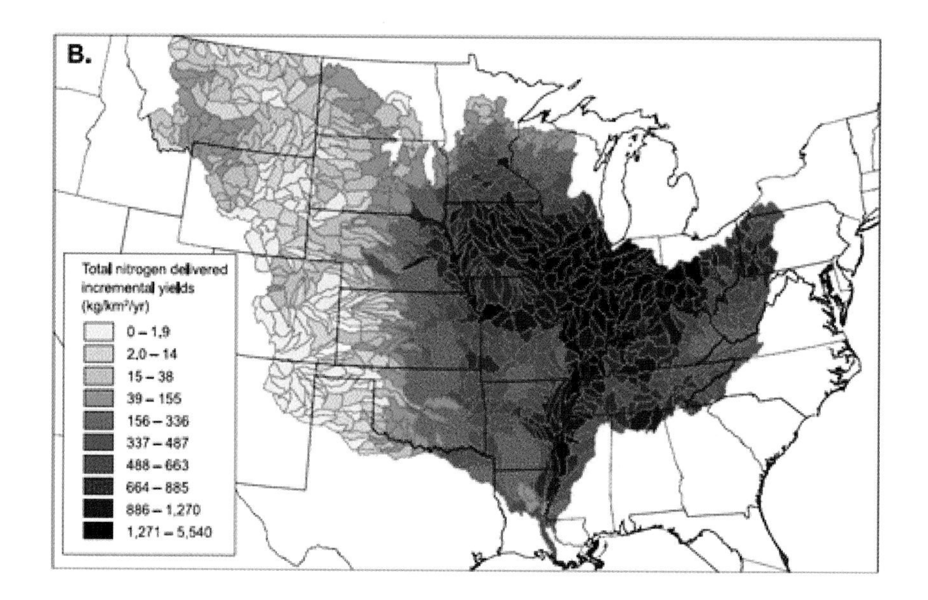

loads delivered to downstream receiving waters (such as the Gulf of Mexico), and demonstrate techniques to place confidence in these rankings. These results will be one of several tools to help guide the allocation of federal funds among States to develop strategies to reduce nutrient loads to the Gulf of Mexico and the Great Lakes.

SPARROW models can be used to estimate how changes in various management decisions should affect water quality and the downstream transport of nutrients, such as with increases in the acreage of corn crops associated with increased ethanol production, decreases in the amount of fertilizers applied to crops, or changes in releases from specific treatment plants.

Regional SPARROW models are being developed in different areas of the country to enable accurate predictions to be made at scales finer than those made with National SPARROW models and are being used to address more regional/local issues.

RESULTS, OUTCOMES OR IMPACTS TO DATE:

SPARROW models have been developed for the Great Lakes/Upper Mississippi River Basins and Entire Mississippi River Basin.

Results of the Mississippi River SPARROW model were used to describe where nutrients (phosphorus and nitrogen) originate from throughout the Mississippi/Atchafalaya River Basin geographically and by land use.

Alexander, R.B., Smith, R.A., Schwarz, G.S., Boyer, E.W., Nolan, J.V., and Brakebill, J.W., 2008, *Differences in phosphorus and nitrogen delivery to the Gulf of Mexico from the Mississippi River Basin:* Environmental Science and Technology, 42(3):822-830.

Results of the Mississippi River SPARROW model were used to describe where nutrients (phosphorus and nitrogen) originate from HUC8 watersheds throughout the Mississippi/Atchafalaya River Basin. The HUC8 watersheds were ranked based on their relative contributions to the Gulf and method of placing certainty in those rankings was developed.

Robertson, Dale M., Schwarz, Gregory E., Saad, David A., and Alexander, Richard B., 2009, *Incorporating Uncertainty into the Ranking of SPARROW Model Nutrient Yields from the Mississippi/Atchafalaya River Basin Watersheds:* Journal of the American Water Resources Association, 45(2):534-549.

PERFORMERS/OTHER PARTNERS (FEDERAL, STATES, OR LOCAL):

U.S. Geological Survey, Wisconsin Water Science Center, Middleton, WI and National Center, Reston, VA.

U.S. Environmental Protection Agency, Region V and Office of Water.

PROJECT PERIOD: 2006-2010

FUNDING LEVELS (CURRENT OR PROPOSED): Funding across years is variable and has been supplied by USGS, NAWQA and the U.S. EPA Office of Water. Four-year funding period includes the approximate sources:

USGS (NAWQA)—$510,000
EPA—$300,000

Additional funding will be required for report preparation for the Mississippi River Basin. Any additional dimensions to this work, such as enhancements to simulate and compare biofuels scenarios, would need additional funding.

NOTABLE EXAMPLES OF FEDERAL RESEARCH ACTIVITIES RELATED TO BIOFUELS AND SUSTAINABILITY

TITLE OF PROJECT OR PROGRAM: Estimated Forest Biomass Supply for the United States–Revision to the Billion Ton Supply Estimates

AGENCY: U.S. Department of Agriculture Forest Service, USDOE Oak Ridge National Laboratory

PROJECT/PROGRAM DESCRIPTION:

The 2005 report "Biomass as a feedstock for a bioenergy and bioproducts industry: The Technical feasibility for a billion ton annual supply" suggested that it may be technically possible to supply up to 1.3 billion tons of wood and agricultural biomass for bioenergy and bioproducts in the United States. This included 368 million tons from wood sources including forest sources, mill residues and urban wastes. Short rotation woody crops were estimated separately as an agricultural source.

It is the objective of a new project to revise these estimates and indicate the economic feasibility of providing forest biomass for bioenergy from each county in the United States Supply curves for forest biomass are being estimated for each county. Forest biomass resources include amounts from current logging residue, amounts from thinnings to mitigate fire hazard and reduce overstocking, amounts from other removals such as land clearing for development, mill residue, urban wood waste from construction and demolition, and conventionally source wood such as pulpwood. The supply curves indicate the amount of wood available at roadside, mill, or urban source at progressively higher costs per oven dry ton.

The estimation effort involves expertise from several disciplines—ecology, silviculture, forest operations and economics. A key concern in estimating amounts from logging residue and thinnings is to assure that the removal amounts are sustainable. Specifically how much logging residue must be left on harvest sites to provide nutrients and habitat? For thinnings, what is the number of years before thinning can recur for each forest type to allow for sustainable regrowth of forests?

Estimates of county level supply curves can be scaled up in at least two ways. Supply curves may be generated for delivery of amounts to any given point by adding transport costs to supply curves from surrounding counties. Supply curves may be added together to estimate state, regional or national level roadside cost supply curves.

Preliminary estimates of forest biomass supply have been used in the report by the Biomass Research and Development Board. These wood biomass supply estimates (along with county level agricultural biomass supply estimates) are being applied/used in the National Biorefinery Siting Project (described separately) to determine the sustainable level of biofuels production in the U.S., and specific biofuels plant locations for optimal production of biofuels given (1) the location of feedstocks, (2) infrastructure to transport feedstock and biofuels and (3) the feedstock demands/costs of conversion technologies. The National Biorefinery Siting Project is funded by USDOE and is being organized by the Western Governors Association. Collaborators include USDA Forest Service, UC Davis, Kansas State University, USDOE Oak Ridge National Laboratory, and others.

RESULTS, OUTCOMES, OR IMPACTS TO DATE:

Forest biomass supply estimates provided for the Biomass Research and Development Initiative report on increasing feedstock production for biofuels suggest forest sources could provide 40 million oven dry tons (odt) per year and produce 4 billion gallons of liquid fuels by 2022. This 40 million odt estimate did not require use of traditionally sourced wood such as pulpwood. However it is likely that pulpwood sources would be used, in part, as demand increased to 40 million odt. It is important to note that the BRDi feedstock estimation project did not consider possible increasing wood biomass demand for electric power production which could increase wood biomass use well beyond 40 million odt per year. In this case it is likely, given our preliminary estimates of wood biomass supply, that notable amounts of conventionally sources wood—pulpwood—would be supplied for biofuels and electric power production.

PERFORMERS/OTHER PARTNERS (FEDERAL, STATES, OR LOCAL):

Members of the team revising the forest biomass supply estimates for the Billion Ton Supply Report.

Ken Skog, Patti Lebow—USDA Forest Service, Forest Products Laboratory, Madison, WI
Marilyn Bufford—USDA Forest Service, Washington, DC
Bryce Stokes—USDOE, Washington, DC (formerly USDA Forest Service, Washington, DC)
Jamie Barbour, Dennis Dykstra—USDA Forest Service, Pacific Northwest Research Station, Portland, OR
Bob Perlack—USDOE Oak Ridge National Laboratory, Oak Ridge, TN

PROJECT PERIOD: 2007-2009

FUNDING LEVELS (CURRENT OR PROPOSED): Forest Service research contributions are funded from annual appropriations.

NOTABLE EXAMPLES OF FEDERAL RESEARCH ACTIVITIES RELATED TO BIOFUELS AND SUSTAINABILITY

TITLE OF PROJECT OR PROGRAM: Discovery Farms

AGENCY: U.S. Geological Survey

PROJECT/PROGRAM DESCRIPTION:

Agriculture has historically been cited as one of the primary causes of water-resource degradation, especially in Wisconsin. Nonetheless, agriculture plays a critical role in the way that we live, the food we eat, and the economics that drive our society. The water-quality implications of a shift toward bio-based energy—whether derived from traditional crops like corn and soybeans, non-traditional crops like switchgrass and woody residues, or manure—can be understood to some degree by extrapolating from our current understandings of the mechanisms by which agricultural practices affect water quality.

Wisconsin producers are facing difficult challenges to remain economically viable: new farm bills are threatening to take away subsidies, increasing fuel and fertilizer costs are limiting profitability, and legislation has been proposed that may significantly change the ways that producers have historically operated. In addition, producers are receiving increased pressure to be "environmentally friendly": well contaminations, manure spills and numerous recent fish kills have all been linked to agriculture.

Bioenergy (in the forms of crop-based biofuels for transportation and biomass-based heat and power) are viewed regionally as an economic opportunity for the Midwest and, nationally, as an environmentally sustainable path to energy independence. At present, there is little empirical evidence to verify these assumptions or guide best practices.

The USGS is cooperating with the Discovery Farms program to collect data to help understand agriculture's impact on the environment and work with producers to evaluate ways to minimize their impact in economically viable ways. The approach is field-based. Monitoring stations installed throughout Wisconsin on selected Discovery Farms represent diverse land characteristics, production schemes, and management styles. Monitoring stations are installed at sites in small, headwater streams, edges of fields, and in subsurface tiles to continuously measure runoff volume during storm-runoff periods, including snowmelt. Samples are combined to represent average concentrations over the duration of a storm; they are analyzed for total phosphorus, dissolved reactive phosphorus, suspended sediment, total dissolved solids, ammonium – N, nitrate + nitrite – N, Kjeldahl – N, and chloride.

RESULTS, OUTCOMES, OR IMPACTS TO DATE:

Two largely overlooked issues that affect agriculture's impact on water quality are weather conditions and the timing of nutrient applications (not just the total amount dictated by a nutrient management plan). The timing, amount, and intensity of rain are HUGE factors in determining runoff of sediment, phosphorus, and nitrogen. Wintertime runoff is particularly important, generating 50 percent or more of total annual runoff. Therefore, the timing of manure applications for fertilizer matters much more than previously understood, no matter what kind of crop is being grown. Manure application to frozen and/or snow-covered ground in February and March is important to water-quality outcomes. During non-frozen ground conditions, especially April through June, water quality is also negatively impacted if runoff occurs before manure is incorporated into the soil. Ranges of sediment, nutrients lost from "typical" Wisconsin fields are 650 pounds of sediment/acre and 2 pounds per acre. Forms of P mostly dissolved P (largely bioavailable) in winter; mostly particulate P in summer.

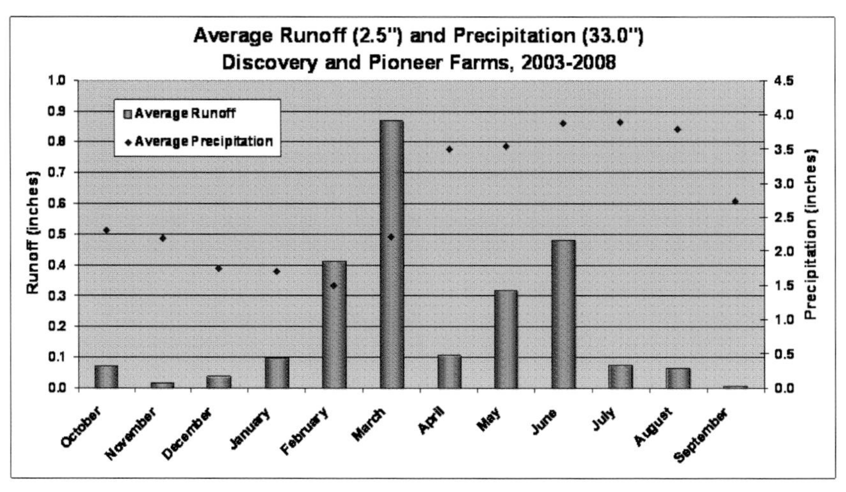

FIGURE 1 Winter runoff of applied manure poses the greatest water-quality risks.

Publications:

Stuntebeck, Todd D.; Komiskey, Matthew J.; Owens, David W.; Hall, David W., 2008, Methods of Data Collection, Sample Processing, and Data Analysis for Edge-of-Field, Streamgaging, Subsurface-Tile, and Meteorological Stations at Discovery Farms and Pioneer Farm in Wisconsin, 2001-7: U.S. Geological Survey Open-File Report 2008-1015, 60 pp.

Komiskey, Matthew J., Stuntebeck, Todd D., Busch, Dennis, Frame, Dennis and Madison, Fred. Nutrients and Sediment in Surface Water Runoff from Frozen Ground Following Manure Applications. Submission pending, 2009.

PERFORMERS/OTHER PARTNERS (FEDERAL, STATES, OR LOCAL):

U.S. Geological Survey, Wisconsin Water Science Center, Middleton, WI
Wisconsin Department of Natural Resources

University of Wisconsin Extension
Sand County Foundation

PROJECT PERIOD: 2001-2010

FUNDING LEVELS (CURRENT OR PROPOSED):

Annual Funding for the USGS portion has ranged between $250K to $350 K for the past five years with the University of Wisconsin Extension Discovery Farms program providing 50 percent, USGS 30 percent, Wisconsin DNR 10 percent, and Sand County Foundation 10 percent No decision has been made to expand to biofuels-related agricultural practices but costs to do so would be in the $150K to $250K range. Overall Discovery Farms annual budget approaches 1 million for all partners.

Appendix D

Brief Survey of State Biofuel Policies in the Upper Midwest

Background Paper for the National Academies' Workshop:
"Expanding Biofuel Production: Sustainability and the Transition to Advanced Biofuels"

June 23-24, 2009
Madison, WI

Bjorn Gangeness
University of Minnesota

INTRODUCTION

Biofuels have been touted as a way to help in achieving energy independence and security, to improve the environment and encourage economic growth in rural areas. Both state and federal governments across the nation have developed policy incentives and other programs to promote the use of bio-based fuels such as ethanol and biodiesel.

Since the late 1970's, the Upper Midwest states have been leaders enacting these policies and supporting programs to encourage the production and use of biofuels building on the region's strong agricultural bases. The evolution and impacts of these policies have implications far beyond the region—to other parts of the United States and the world. New policies will likely be necessary to address emerging questions about the economic, environmental, and social impacts of current biofuels and to support a transition to more sustainable advanced biofuels.

U.S. BIOFUEL PRODUCTION HISTORY

Biofuel production has grown dramatically in the last thirty years. In 1980 the United States consumed only 83 million gallons of ethanol in vehicle fuel, and by 2007 this number had reached almost 6.8 billion gallons (Figure 1). Biodiesel consumption was about 9 million gallons in 2001(when statistics were first published) and increased to almost 0.5 billion gallons in 2007.[1]

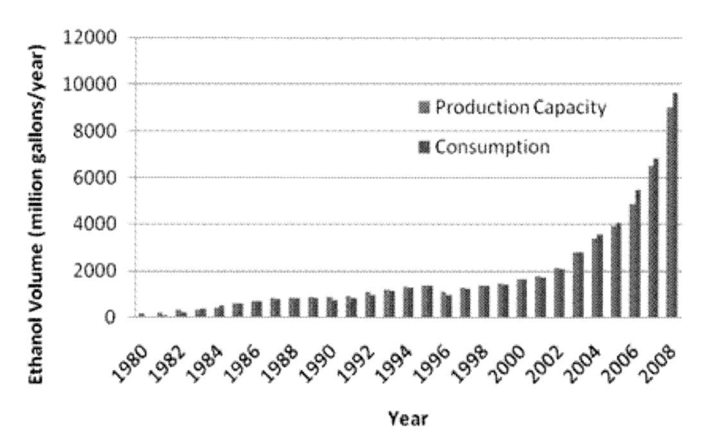

FIGURE 1 U.S. ethanol vonsumption and production capacity 1980-2008.

Market gains of this magnitude are significant but the promising future originally imagined by biofuels' biggest proponents has not materialized, caught, in part, by declines in world petroleum prices and the general economic downturn. At the same time, larger questions of environmental sustainability and economic efficiency have emerged. Advanced ligno-cellulosic biofuels derived from more complex organic feedstocks like wood or grasses are being promoted as more environmentally beneficial than so-called first-generation biofuels and are seen as a way to avoid impacting food supplies and prices. Aggressive goals have been set for advanced biofuels production to be met by 2015. But the transition to large scale commercial advanced biofuels production is still some years away.[2]

POLICY IMPLICATIONS OF BIOFUELS

Policies to promote the use of biofuels have lowered the cost to producers of entering the market and encouraged increased consumption of biofuels as an

[1]Fuel Ethanol and Biodiesel Overview, 1981-2007. Accessed May 22, 2009. *http://www.eia.doe.gov/emeu/aer/txt/ptb1003.html.*

[2]U.S. Energy Policy Act of 2005 (EPAct 2005) Accessed May 22, 2009. *http://www.epa.gov/oust/fedlaws/publ_109-058.pdf.*

alternative or additive to petroleum based fuels. These policies various types of subsidies, tax incentives, mandates, and investment credits.

In 2006 there were over 150 policies explicitly supporting biofuels at the state level and another 30+ policies at the federal level.[3] The aggregate affect of this abundance of support has been to increase the production and consumption of biofuel products. However, the ultimate effect on the economy and the environment from increased production and consumption of biofuels is unclear.

BIOFUELS DEPLOYMENT IN IOWA, MINNESOTA, AND WISCONSIN

The Upper Midwest has emerged as the largest source of ethanol production in the country. Three states in particular have embraced the potential benefits of biofuels and deliberately bolstered support of biofuel production over the past 30 years. This support has led to dramatic increases in the acreage devoted to biofuel feedstocks—primarily corn—and to the construction of ethanol plants with a large total capacity. Iowa, Minnesota and Wisconsin together have 35 percent of the total U.S. nameplate capacity for producing ethanol.[4] Iowa alone delivers almost 25 percent of U.S. ethanol production (Figure 2). Biodiesel production is more nationally diffuse. These three states have a total biodiesel production capacity of 376.5 million gallons or about 15 percent of national capacity.[5]

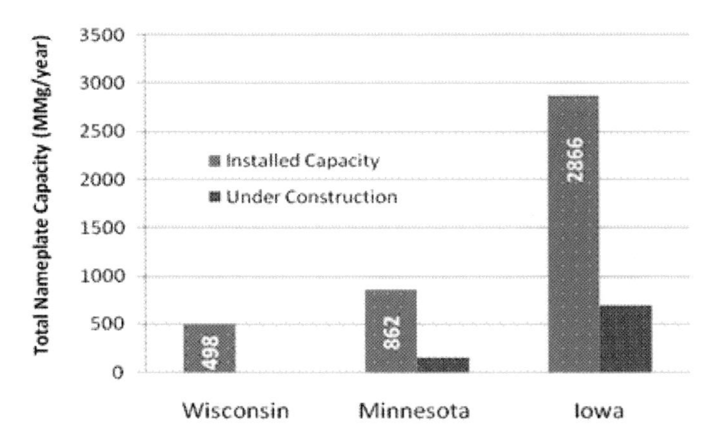

FIGURE 2 Ethanol production capacity.

Each state has taken a different approach to biofuels development. In 1992, Minnesota passed a law requiring that all gasoline sold in the state to be blended

[3]Koplow, D. 2006. "Biofuels—At What Cost ?: Government Support for Ethanol and Biodiesel in the United States". International Institute for Sustainable Development.

[4]Nebraska Energy Office. Accessed May 26, 2009. *http://www.neo.ne.gov/statshtml/122.htm.*

[5]National Biodiesel Board. Accessed May 26, 2009. *http://www.biodiesel.org/.*

with 10 percent ethanol increasing to 20 percent in 2013.[6] Wisconsin has considered a renewable fuel standard (RFS) but so far has only required state fleet vehicles to use at least 10 percent ethanol blended gasoline since 1993.[7] Iowa has mandated that vehicles sold in-state be operable on gasoline blended with at least 10 percent ethanol or more since 1993 and recently passed a RFS mandating 10 percent renewable blends in all gasoline starting in 2009 increasing to 25 percent in 2020.[8]

STATE BIOFUELS POLICY: DRIVERS AND TOOLS

Originally, biofuels policies like the Minnesota blending requirement were introduced by state legislators who saw significant direct benefits to their local constituents or customers. More recently, governors of these states have played a bigger role by proposing renewable fuel mandates like Minnesota's biodiesel blending mandate of B2 (2 percent biodiesel in diesel fuel) by 2005 increasing up to B20 by 2015. Wisconsin Governor Jim Doyle recently ordered a reduction of 20 percent in petroleum-based content consumed in state gasoline-fueled vehicles by 2010 and a 50 percent reduction by 2015.[9] In the case of biofuel development, financial incentives to promote biofuel production within state borders include:

- elimination or reduction of excise taxes (currently federal);
- renewable fuels standards
- accelerated capital depreciation tax benefits;
- alternative fuel vehicle mandates;
- state fleet fuel consumption quotas;
- loan-guarantee programs;
- producer payments;
- feedstock subsidies;
- blender's credits;
- investment and production tax credits (ITCs & PTCs);

The Upper Midwest uses all of these incentives[10] to promote biofuel production and use (Table 1).

---

[6]Brown, E., K. Cory, and D. Arent. January 2007. "Understanding and Informing the Policy Environment: State-Level Renewable Fuels Standards." NREL Technical Report.

[7]Wisconsin Biofuels and Alternative Fuels Use Report 2007 Annual Report. Accessed May 26, 2009. *http://energyindependence.wi.gov/docview.asp?docid=11267&locid=160.*

[8]Iowa Office of Energy Independence. Accessed May 26, 2009. *http://www.energy.iowa.gov/* renewable_fuels/*IA_renewfuels_standard.html.*

[9]Wisconsin Biofuels and Alternative Fuels Use Report 2007 Annual Report. Accessed May 26, 2009. *http://energyindependence.wi.gov/docview.asp?docid=11267&locid=160.*

[10]Koplow, D. 2006. "Biofuels—At What Cost ?: Government Support for Ethanol and Biodiesel in the United States". International Institute for Sustainable Development.

TABLE 1 State Programs Supporting Biofuel Production and Use

Policy/State	Iowa	Minnesota	Wisconsin
Renewable Fuel Standard (RFS)	Equivalent of 25 percent of all gasoline sales must come from renewable sources [either 10 percent or 85 percent ethanol blends (E10, E85), or biofuel that is 1 percent biodiesel by volume (B1) at a minimum] (2006) (NREL 2007)[a]	Total motor gasoline sales must contain 20 percent ethanol by volume (E20) by 1/2013. Total diesel sales must contain 5 percent biodiesel by volume (B5) by 5/2009. (1991, revised 2006) (MN Next Generation Energy Act 2007)[b]	
Retail Infrastructure Incentives	50 percent+ cost-share program for E85 and biodiesel (B1+) dispensers (2005-2008). Iowa Power Fund grants and low-interest loans. (Iowa Office of Energy Independence presentation 2009)[c]	50 percent Cost share for E85 fueling stations up to $15,000 (MN Office of Energy Security 2009)[d]	Ethanol Refueling Project has provided $100,000 to assist with construction of E85 fueling stations (leveraged DOE funding) (WI Office of Energy Independence 2007)[e]
Retail tax credit	25 cpg (cents per gallon) tax credit to distributors of E85, 3 cpg to retailers selling B2 or higher and more than 50 percent of total sales (Koplow 2006)[f]		
State-backed Loan Programs	Low to 0 percent interest loans backed by up to 50 percent public funding to renewable energy production facilities up to max of $250,000 (Koplow 2006)	Low-interest state loans from the Minnesota Investment Fund (MN OLA 2009)[g]	
Producer payments		20 cpg ethanol for the first 15 million gallons of annual production (1987-2012) (MN OLA 2009)	

continued

TABLE 1 Continued

Policy/State	Iowa	Minnesota	Wisconsin
Ethanol Blend or Renewable Content Goals (not RFS)			Executive Order in 2006 requiring state vehicles to reduce fossil fuel use (20 percent by 2010 and 50 percent by 2015) and replace it with increased use of E10, E85, and biodiesel. Also, governor set a goal of 25 percent renewable power and fuel in WI by 2025 (WI Office of Energy Independence 2007)
Blender's Credit		Payment to blenders peaking at a total of $25 million/year in 1994 (1980-mid 1990s) (MN OLA 2009)	
Production or Investment Tax Credit	General business development investment tax credit programs that are used significantly by biofuel production facilities. Usually targeted specifically to job creation or property tax relief. (Iowa Dept of Revenue 2009)[h]		PTC of 20 cpg on first 15 MGY of ethanol production; capped at $3 million over 5 years. Expired 1 July 2006 (Koplow 2006)

[a]Brown, E., K. Cory, and D. Arent. 2007. NREL Technical Report. "Understanding and Informing the Policy Environment: State-Level Renewable Fuels Standards." *http://www.osti.gov/bridge/product.biblio.jsp?query_id=1&page=0&osti_id=898863.*

[b]Minnesota Next Generation Energy Act of 2007. Accessed May 30, 2009. *http://www.nextstep.state.mn.us/res_detail.cfm?id=4034.*

[c]Iowa Office of Energy Independence. Accessed May 26, 2009. *http://www.energy.iowa.gov/renewable_fuels/IA_renewfuels_standard.html*

[d]Minnesota Department of Commerce–Office of Energy Security. "Renewable and Efficiency Incentives." Accessed May 30, 2009 *http://www.state.mn.us/portal/mn/jsp/content.do?id=-536893811&contentid=536885915&contenttype=EDITORIAL&programid=536917180&agency=Energy.*

[e]Wisconsin Biofuels and Alternative Fuels Use Report 2007 Annual Report. Accessed May 26, 2009. *http://energyindependence.wi.gov/docview.asp?docid=11267&locid=160.*

[f]Koplow, D. 2006. "Biofuels—At What Cost ?: Government Support for Ethanol and Biodiesel in the United States". International Institute for Sustainable Development.

[g]Office of the Legislative Auditor. 2009. "Biofuel Policies and Programs." State of Minnesota. *http://www.auditor.leg.state.mn.us.*

[h]Jin, Z. and B. Teahan. 2009. "Iowa's Tax Incentive Programs Used by Biofuel Producers Tax Credits Program Evaluation Study." Iowa Department of Revenue. *http://mpra.ub.uni-muenchen.de/14795/.*

BIOFUELS POLICIES WITH SIGNIFICANT IMPACTS

The States of Iowa, Minnesota, and Wisconsin have each taken a somewhat different path in supporting the development of their biofuel economies. For example, Minnesota introduced a ethanol blender's tax credit of 4 cpg to support the blending of ethanol in gasoline in 1980 and a 20 cpg producer payment in 1986. Iowa and Minnesota both established renewable fuels standards requiring increasing percentages of ethanol and biodiesel contained in the states' transportation fuels. Wisconsin has used a combination of a very successful producer payment program with other diverse tax credits and renewable energy loan programs to cultivate its biofuel economy.

Minnesota

In addition to the tax credits and producer payments, Minnesota has supported ethanol and biodiesel development with many other incentives and programs. The cost-share program for fueling station installation helps to increase the availability of E85 at fuel stations across the state. State managed low-interest loans were used for some biofuel production facilities, but new facilities are now usually subsidized by an economic development program called JOBZ that is not biofuel specific but provides tax relief to new businesses locating in rural areas.[11] It is increasingly common for states to support biofuels development with economic development initiatives that are not specific to biofuels.

Iowa

Iowa's biofuel industry has recently taken advantage of economic development incentives included in the Enterprise Zone and High-Quality Job Creation Programs.[12] Ethanol plants have been awarded over $405 million worth of tax credits through these programs. Iowa now also has a Renewable Fuels Standard beginning in 2009 with 10 percent of gasoline sales must be renewably sourced (i. e., corn or other renewable feedstock). This percentage must rise to 25 percent renewably sourced by January 1, 2021 according to the RFS.[13] A tax credit measured by pure ethanol gallons sold is also defined in the RFS if retailers reach a certain percentage of sales from ethanol (3 percent below threshold rising to full credit at the threshold point of 10 percent pure ethanol sales in 2009).

---

[11] Office of the Legislative Auditor. 2009. "Biofuel Policies and Programs." State of Minnesota. *http://www.auditor.leg.state.mn.us.*

[12] Jin, Z. and B. Teahan. 2009. "Iowa's Tax Incentive Programs Used by Biofuel Producers Tax Credits Program Evaluation Study." Iowa Department of Revenue. *http://mpra.ub.uni-muenchen.de/14795/.*

[13] *http://www.energy.iowa.gov/renewable_fuels/ethanol/retailers_producers.html.*

Wisconsin

In contrast to the tax credits in Minnesota and Iowa, Wisconsin has primarily relied upon grants and loans to stimulate the production and distribution of biofuels. One exception is the ethanol production tax credit that was in place for 2 years and was considered so successful that the program was retired after many facilities came online. The governor has also used Executive Order numbers 141 and 192 to promote the use of renewable fuels in state fleet vehicles and to require the use of 25 percent renewably sourced energy for power and transportation by the year 2025. One loan program, called the Freight Railroad Infrastructure Improvement Program is intended to expand freight capacity for biofuels and other products, has distributed $13.5 million for ethanol plants. The Wisconsin Energy Independence Fund has awarded $7.5 million 50 percent cost share grants to renewable energy projects in 2008 alone.[14]

Midwest Governors Association

In addition to state-specific initiatives, the Midwest Governors Association has also been active in promoting renewable energy and climate change mitigation policies. Though the MGA does not write or implement policy for particular states, it often advocates for positions to the U.S. Environmental Protection Agency and lobbies Congressional leaders for action on energy-related policies at the federal level. The latest MGA letter to EPA called for ethanol blending limits to be raised to 15 percent.[15] In 2007 the MGA also released recommendations to establish a regional greenhouse gas management initiative that would rely on biofuels to reduce greenhouse gas emissions, among other actions.

ADVANCED BIOFUEL PRODUCTION

Lawmakers are beginning to turn to advanced biofuel production as a potential way to avoid the adverse environmental consequences often seen with first generation biofuels and to assure that biofuels production does not compete with food production. Advanced, or second generation, biofuels are produced from biomass that is high in lignin or contains high levels of cellulose, such as trees, shrubs, grasses, or corn-stover. These materials cannot be processed into ethanol or biodiesel by the same technology as corn or soybeans. Lignin and cellulose must first be broken down into the simple sugars and oil that can then be processed into ethanol or biodiesel. This initial step has proven to be expensive

[14]Presentation given at National Academies workshop on "Sustainability and the Transition to Advanced Biofuels." June 23, 2009. Madison, WI. *http://sustainability.nationalacademies.org/pdfs/Ziewacz.pdf.*

[15]*http://www.midwesterngovernors.org/.*

but public and private research institutions continue to invest heavily in finding a way over this hurdle.[16]

ADVANCED BIOFUEL DEVELOPMENT PROGRAMS

There are many programs that aim to make cellulosic biofuels commercially viable. Most of these programs are set by federal policies including the 2008 Farm Bill, the Energy Policy Act of 2005, and the Energy Independence and Security Act (EISA) of 2007.[17] These programs are primarily directed at research and development administered through the Department of Energy or the U.S. Department of Agriculture. The EISA 2007 mandates the use of 950 million gallons of biomass-based fuel in the United States in 2010.[18] Some advanced biofuels policies have been proposed or passed at the state level, but at substantially lower levels of support. In 2007, Minnesota began creating a program called Reinvest in Minnesota—Clean Energy that proposed using land easements to support the conversion of agricultural land into dedicated cellulosic energy crops. New York has supported the construction of cellulosic ethanol pilot facilities.[19] Due to the nascent development of these advanced biofuels, few significant state incentives are currently in place.

FUTURE CHALLENGES

Economic—Biofuels have the potential to reduce U.S. dependency on oil. However, studies to date suggest that the ability to reduce U.S. oil imports will not be as dramatic as proponents have claimed. Additional economic considerations include indirect consequences of public subsidies to farmers and producers. Subsidies provided to farmers for the production of agricultural commodities—including biofuel feedstocks—continue to be the subject of international trade debates. In the case of U.S. corn production it is estimated that ethanol accounted for 12 percent of the crop in 2005 increasing to a projected 23 percent in 2014-2015.[20] In the US, crop subsidy payments vary with market conditions but averaged about $5 billion/year for 2000-2004. It is also estimated that ethanol captured nearly $1 billion of this annual average and that percentage would increase if more corn was used for fuel. If corn continues to be the

---

[16]Gardner, Tim. "US drafts rule to lower CO2 output from biofuels." May 5, 2009. *http://www.reuters.com/article/idUSTRE5443L320090505.*

[17]Department of Energy Office of Science. "Biofuels Policy and Legislation." Accessed May 22, 2009. *http://genomicsgtl.energy.gov/biofuels/legislation.shtml.*

[18]Department of Energy Office of Energy Efficiency and Renewable Energy. "Federal Biomass Policy." Accessed May 22, 2009. *http://www1.eere.energy.gov/biomass/federal_biomass.html.*

[19]Koplow, D. 2006. "Biofuels—At What Cost ?: Government Support for Ethanol and Biodiesel in the United States". International Institute for Sustainable Development.

[20]Baker, A and S Zahniser. April 2006. "Ethanol Reshapes the Corn Market." Amber Waves 4, no. 2 pp. 30-35.

main feedstock for ethanol production and crop yields do not keep up with this growth, other areas of the economy are likely to be affected with rising food and fuel prices.[21]

Environmental—Advanced biofuels have the potential to improve upon the environmental impacts associate with the production and use of first generation of biofuels by reducing the need for fertilizers and pesticides, reducing water use, and eliminating competition between crops for food and fuel. Supporting the development of the market for advanced biofuels will be important to their success. If policy makers decide that achieving improvements on these impacts is important, the environmental, economic and social impacts must be considered in crafting appropriate advanced biofuels policies.

ANNEX D

STATE POLICIES AND PROGRAMS INFLUENCING BIOFUELS DEVELOPMENT

Minnesota

Tax Credits

A. Blender's Credit
4 cent per gallon tax credit to blenders distributing gasoline with 10 percent ethanol. Program credits peaked at a total of $25 million/year in 1994 (1980–mid-1990s) (MN OLA 2009)[22]

B. JOBZ Business Tax Credit
8 ethanol facilities and 2 biodiesel facilities have benefited from corporate, property, and sales-tax credits through this program. 370 other non-biofuels industries have also benefited, totaling over $78 million in tax credits. (MN Department of Employment and Economic Development 2009)[23]

Grants and Loans

C. Retail Infrastructure Incentives
50 percent Cost-share for E85 fueling stations up to $15,000 (MN Office of Energy Security 2009) [24]

[21]Runge, CF and B Senauer. "How biofuels could starve the poor." Foreign Affairs. May 2007. *http://www.foreignaffairs.org/20070501faessay86305-p20/c-ford-runge-benjamin-senauer/how-biofuels-could-starve-the-poor.html.*

[22]Office of the Legislative Auditor. 2009. "Biofuel Policies and Programs." State of Minnesota. *http://www.auditor.leg.state.mn.us.*

[23]*http://www.deed.state.mn.us/bizdev/jobz.htm.*

[24]*http://www.house.leg.state.mn.us/hrd/pubs/ss/ssethnl.htm.*

D. Producer payments
20 cent per gallon ethanol for the first 15 million gallons of annual production, limited to $3 million/year per facility (1987-2012) (MN OLA 2009)

Executive Orders or Legislative Mandates

E. Renewable Fuel Standard (RFS)
Total motor gasoline sales must contain 20 percent ethanol by volume (E20) by 2013. Total diesel sales must contain 5 percent biodiesel by volume (B5) by 5/2009. (1991, revised 2006) (MN Statutes 239 sections 77 and 791)

Advanced Biofuels Policies

F. Next Generation Energy Board grants
A one-time appropriation of $3 million to be granted to renewable energy projects, including advanced biofuels. 2 of 8 grants went to existing biofuels projects to study feasibility of co-locating cellulosic ethanol with conventional ethanol and to convert another ethanol project to be powered by biomass instead of natural gas. (MN OLA 2009)

G. Reinvest in Minnesota Clean Energy
A proposed agricultural easement program focused on supporting a conversion to perennial energy crops. Level of payment considered at 80 percent of the value of the land converted for a 20 year or longer contract. (proposed but defeated legislation 2008)(Board of Water and Soil Resources 2008)[25]

Iowa

Tax Credits

A. Retail tax credit
25 cent per gallon (cents per gallon) tax credit to distributors of E85 dropping to 20 cpg for 2009-2010, 3 cpg to retailers selling B2 or higher and more than 50 percent of total sales (Iowa Office of Energy Independence 2009)[26]

B. Investment Tax Credit
Iowa has a few general business development investment tax credit programs that are used significantly by biofuel production facilities. Usually targeted specifically to job creation or property tax relief. Some programs have minimum wage requirements. More than $405 million

[25] *http://www.bwsr.state.mn.us/RIM-CE.html.*
[26] *http://www.energy.iowa.gov/renewable_fuels/ethanol/tax_changes.html.*

were claimed in tax credits for a total of 55 ethanol projects through 2008 (Iowa Department of Revenue 2009)[27]

C. Ethanol Promotion Tax Credit
The incentive each fuel retailer is eligible for will be directly tied to the preceding IRFS schedule with amounts to be paid only for "pure" ethanol gallons (E100). The total credit will be based on how closely the fuel sales at each site achieve the IRFS and how many gallons are sold. The credits range from 2.5 to 6.5 cent per gallon depending on the threshold and year until 2020. (Iowa Office of Energy Independence 2009)

D. Biodiesel Blended Fuel Tax Credit
A retail dealer who sells or dispenses biodiesel blended fuel is eligible for this new income tax credit. The biodiesel must be at least 2 percent blended for the tax credit to apply. The retailer qualifies for the tax credit if they sell 50 percent or more of at least B2 biodiesel from the entire volume of diesel they sell, and the 3 cents per gallon tax credit will apply to every gallon of biodiesel sold if they qualify. (Iowa Office of Energy Independence 2009)

Grants and Loans

E. Iowa Power Fund
$100 million appropriated over 4 years. Provides financial assistance to projects that will improve Iowa's biofuels, renewable energy, and energy efficiency sectors. The Fund's board is comprised of 18 members who review and approve project applications. (Iowa Office of Energy Independence 2009)[28]

F. Retail Infrastructure Incentives
50 percent+ cost-share program for E85 and biodiesel (B1+) dispensers (2005-2008). (Iowa Renewable Infrastructure Board 2009)[29]

G. Alternate Energy Revolving Loan Program (AERLP)
The revolving loan funds equal to 50 percent of the total financed cost of a project (up to $1 million) at 0 percent interest. The AERLP has served a balanced mix of technologies including solar, biomass, small hydro and small and large wind turbine facilities since 1997. (Iowa Energy Center at Iowa State University 2009)[30]

[27]Jin, Z. and B. Teahan. 2009. "Iowa's Tax Incentive Programs Used by Biofuel Producers Tax Credits Program Evaluation Study." Iowa Department of Revenue. *http://mpra.ub.uni-muenchen.de/14795/.*

[28]*http://www.energy.iowa.gov/Power_Fund/about_IPF.html.*

[29]*http://www.iowalifechanging.com/business/renewablefuels.aspx.*

[30]*http://www.energy.iastate.edu/AERLP/index.htm.*

Executive Orders or Legislative Mandates

H. Iowa Renewable Fuel Standard (RFS)
Equivalent of 25 percent of all gasoline sales must come from renewable sources [either 10 percent or 85 percent ethanol blends (E10, E85), or biofuel that is 1 percent biodiesel by volume (B1) at a minimum] (2006) (NREL 2007)[31]

I. Executive Order 3 (6/21/2007) from Governor Culver ordered the use of E-85 fuel in the state's flexible-fuel vehicles shall be increased to at least 60 percent of fuel purchases, and instructs the Office of Energy Independence and DAS to develop a "State Government E-85 Use Plan. (State Library of Iowa 2009)[32]

J. Executive Order 6 (2/21/08) from Governor Culver ordered the creation of a Biofuels Task Force to oversee the increased use of biofuels in state vehicles, decreased miles traveled, and higher energy efficiency in the state vehicle fleet, among other environmental actions. (State Library of Iowa 2009)[33]

K. Executive Order 16 (8/20/2009) from Governor Culver ordered the creation of the Iowa Green Jobs Task Force. The fifteen member task force is to help focus state government's efforts in creating high-paying, green-collar jobs, as well as coordinate the state's efforts to secure federal green initiative grants through the American Recovery and Reinvestment Act. (State Library of Iowa 2009)[34]

Wisconsin

Tax Credits

A. Production Tax Credit for Ethanol
20 cent per gallon production tax credit on first 15 MGY of ethanol production; capped at $3 million over 5 years. Expired 1 July 2006 [s. 560.031] (Koplow 2006)

B. Production Tax Credit for Biodiesel
10 cent per gallon production tax credit on biodiesel, 2.5 million gallon minimum up to $1 million for tax years 2010 through 2012 [s. 71.07 (3h)][35] (*pending legislation*)

---

[31] *http://www.energy.iowa.gov/renewable_fuels/IA_renewfuels_standard.html.*
[32] *http://publications.iowa.gov/5190/1/03-070621.pdf.*
[33] *http://publications.iowa.gov/6275/1/06-080221 percent5B1 percent5D.pdf.*
[34] *http://publications.iowa.gov/7949/1/Executive_Order_No16.pdf.*
[35] *http://www.legis.state.wi.us/RSB/STATS.HTML.*

C. Ethanol and Biodiesel Fuel Pump Income Tax Credit
Tax credit for 25 percent of the cost of installation, up to $5,000 for tax years 2007 thru 2017. [Statute s. 71.07 (5j)](WI Office of Energy Independence 2009)

D. Business Tax Credit
Full income and franchise tax credit for motor vehicles that use gasoline and ethanol mixtures as fuel and for fuel efficient hybrid motor vehicles up to $1000 per year. SB-138[36] (*pending 2009 legislation*)

Grants and Loans

E. Energy Independence Fund
Renewable energy grants and loans for up to 50 percent of project costs. $7.5 million awarded in 2008 totaling 22 projects. (WI Office of Energy Independence 2009)[37]

F. Agricultural Diversification Program
Supports agriculture-related projects with grants of 75 percent of projects cost up to $50,000. 20 grants have been awarded totaling $600,000 since 2002. [s. 93.46 (1), (2), and (3)] (WI Office of Energy Independence 2009)

G. Biogrant Program
Funding provided to 12 projects since 2006. $677,000 has supported 7 biofuels projects. (WI Office of Energy Independence 2009)

Transportation Facilities Economic Assistance and Development.
50 percent cost-share grants to biofuels facilities based on job creation. 5 facilities have been funded totaling $665,400. (WI Office of Energy Independence 2009)

H. Freight Railroad Infrastructure Improvement Loan program.
100 percent loans to connect industry to rail or rail improvements repaid from 2 percent up to the prime interest rate. 4 biofuels facilities have benefited from the program for a total of $13.5 million. (WI Office of Energy Independence 2009)

I. Ethanol Refueling Project
$100,000 in total assistance provided for construction of E85 fueling stations (leveraged DOE funding) (WI Office of Energy Independence 2007)[38]

---

[36] *http://www.legis.state.wi.us/2009/data/SB138hst.html.*
[37] *http://energyindependence.wi.gov/category.asp?linkcatid=2991&linkid=1462&locid=160.*
[38] *http://www.wisgov.state.wi.us/journal_media_detail.asp?prid=1554.*

Executive Orders or Legislative Mandates

J. Executive Order 141 (2006)
Requires state vehicles to reduce fossil fuel use (20 percent by 2010 and 50 percent by 2015) and replace it with increased use of E10, E85, and biodiesel. Also, governor set a goal of 25 percent renewable power and fuel in WI by 2025 (WI Office of Energy Independence 2007)[39]

Advanced Biofuels Policies

K. Extensive advanced biofuels legislation introduced in the Assembly providing financial assistance to the use of bioenergy feedstocks, biorefineries, and biomass energy, and providing some tax credits for the use of renewable fuels. AB-408/SB-279[40] (Office of Energy Independence 2009, personal communication)

[39] *http://www.wisgov.state.wi.us/journal_media_detail.asp?locid=19&prid=2588.*
[40] *http://www.legis.state.wi.us/2009/data/SB279hst.html.*

Appendix E

Assessing the Sustainability of Biofuels: Metrics, Models, and Tools for Evaluating the Impact of Biofuels

Background Paper for the National Academies' Workshop:
"Expanding Biofuel Production: Sustainability and the Transition to Advanced Biofuels"

June 23-24, 2009
Madison, WI

Chris Tessum, Adam Boies, Jason Hill, and Julian D. Marshall
University of Minnesota

INTRODUCTION

This background paper explains and discusses concepts and issues related to the sustainability of biofuels, including the definition of sustainability in general and as related to biofuel production, the proposed and implemented regulatory frameworks aimed at labeling and controlling the sustainability of biofuel production, and the software tools available to quantify various aspects of sustainability.

SUSTAINABILITY FRAMEWORKS

The use of the term "sustainability" is so widespread in the discussion of anthropogenic impacts on our planet that its meaning in several contexts, including biofuels, is ill defined. Multiple sustainability frameworks are available, most or all of which are applicable to biofuel production:

The Triple Bottom Line. This framework, also called the "3 E's" (environment, equity, economy), holds that to be sustainable an organization must consider the environmental and social aspects of its actions as well as economic returns.

The Natural Step. The Natural Step defines a sustainable society as one in which "nature is not subject to systematically increasing (1) concentrations of substances extracted from the earth's crust, (2) concentrations of substances produced by society, or (3) degradation by physical means; and, in that society, (4) human needs are met worldwide" (Nattrass and Altomare, 1999).

The Ecological Footprint. Here, sustainability, defined as "living within the regenerative capacity of the biosphere" (Rees and Wackernagel, 1994; Wackernagel et al., 2002), involves comparing the amount of land required to produce food and other goods for, and to absorb wastes from, society to the amount of land available. Wackernagel et al. (2002) calculated that human demand may have been in excess of the Earth's regenerative capacity since the 1980s, and is currently 20 percent above capacity.

Graedel and Klee's Sustainable Emissions and Resource Usage. Graedel and Klee (2002) quantify a sustainable activity in the following steps: (1) Establish the available supply or limit of the chosen resource or product. (2) Choose a time period over which the use of the resource or creation of the product cannot exceed the supply or limit (e.g., 50 years). (3) Account for recapture (e.g., recycling, sequestration). (4) Using this information, derive the maximum acceptable rate of use or production and compare it to the current rate. If the current rate is higher than the maximum rate, it is unsustainable.

Marshall-Toffel Sustainability Hierarchy. Marshall and Toffel (2005) review previous frameworks, and then prioritize sustainability goals into a four-level hierarchy, starting with the most important as follows: "(1) Actions that, if continued at the current forecasted rate, endanger the survival of humans. (2) Actions that significantly reduce life expectancy or other basic health indicators. (3) Actions that may cause species extinction or violate human rights. (4) Actions that reduce quality of life or are inconsistent with other values, beliefs, or aesthetic preferences."

DISCUSSION

As Marshall and Toffel (2005) highlight, each of the frameworks listed above has strengths and weaknesses. For example, the Triple Bottom Line provides a method for corporations to increase their sustainability, but critics point out that it is arbitrary to stop at three constraints when other factors (e.g., ethics) are also important. The social and environmental performance of a company can also be difficult to quantify.

A strength of the Natural Step is that it presents quantifiable indicators for sustainability. However, it does not address the relative importance of specific criteria. Also, it is difficult to relate the Natural Step criteria to physical effects of unsustainability. For example, a decrease in tropospheric carbon dioxide (CO_2) concentration from 380 to 360 parts per million (ppm) could be considered sustainable by the definition above, yet those concentrations are still elevated above the historical average (280 ppm), and would likely continue anthropogenic climate change.

The Ecological Footprint can illustrate the relative sustainability of different practices by calculating how the footprint would change if all of society adopted a given practice. However, the data required to calculate an Ecological Footprint are difficult to obtain, and difficult to update to account for improvements in technology.

The Graedel and Klee method is novel because it introduces the need for a time scale of sustainability, and the idea that nonrenewable resource use can be sustainable up to a certain rate. The method is only applicable to single resources or products, so the application of this method to a suite of resources, products, and limitations, such as those necessary for biofuel production, would require multiple analyses.

The sustainability hierarchy (Marshall and Toffel, 2005) attempts to combine and prioritize aspects of the frameworks listed above it. The first three levels of the hierarchy are readily quantifiable within the current scope of scientific inquiry. (Marshall and Toffel argue that the fourth level should not be included in the definition of sustainability because values, beliefs, and aesthetic preferences vary among people and cultures and change over time.) Because the sustainability hierarchy encompasses goals from the frameworks reviewed here, it will be used to review the strengths and weaknesses of the extant principles for biofuel sustainability. It should be noted that whether level 1, 2, or 3 applies to a certain situation depends on the severity of the situation, among other factors. For instance, severe levels of emissions of greenhouse gases could cause the extinction of the human species (level 1), while less severe emission levels may only reduce human life expectancy (level 2). In this report, sustainability principles are split into two categories: levels 1-3 versus level 4 from the hierarchy.

BIOFUEL SUSTAINABILITY PRINCIPLES, CRITERIA, AND INDICATORS

To determine whether individual instances of biofuel production are sustainable within the general frameworks above, principles, criteria, and indicators have been developed. *Principles* are general tenets that adapt the sustainability frameworks discussed above for biofuel production. *Criteria* are conditions to be met to achieve these tenets. *Indicators* are measurable tests to determine whether individual farms, producers, or companies are meeting the criteria. Some examples of frameworks for principles, criteria, and indicators involving the United States follow.

Principles

Roundtable on Sustainable Biofuels Standard

The Roundtable on Sustainable Biofuels (RSB) standard is an international, multiparty attempt to define the requirements for sustainable biofuel production. The current version of the standard is version zero. Table 1 summarizes the RSB principles and the applicable levels of the Marshall-Toffel Sustainability Hierarchy.

Principles 3, 4, 6, 7, 8, 9, and 10 of the RSB standard all relate directly to either human health or survival, human rights, or species extinction; therefore, therefore, they are included in the first three levels of the sustainability hierarchy. Principles 1, 2, 5, and 12 relate to quality of life or values and beliefs; therefore, they fall into the fourth level of the hierarchy. Principle 11, cost-effectiveness, does not fit within the hierarchy, but it is clearly a consideration for any biofuel.

25x25 Sustainability Principles

The Energy Independence and Security Act of 2007 (EISA) established a U.S. goal to derive 25 percent of U.S. energy use from renewable sources by 2025. The related action plan specified sustainability as one of the main requirements for successful realization of the Act. It defined sustainability as "…[To] conserve, enhance, and protect natural resources and be economically viable, environmentally sound, and socially acceptable." To encourage sustainable biomass production, EISA developed the principles summarized in Table 2.

TABLE 1 Summary of Roundtable for Sustainable Biofuels Sustainability Principles and Applicable Levels of the Marshall-Toffel Sustainability Hierarchy

Hierarchy	Principles	Levels 1-3	Level 4
1	Obey all local laws and international treaties.		X
2	Consider all relevant stakeholders.		X
3	Reduce greenhouse gas emissions relative to fossil fuels.	X	
4	Obey all human rights and worker rights.	X	
5	Contribute to rural development.		X
6	Do not impair food security.	X	
7	Avoid negative impacts on biodiversity, ecosystems, and areas of high conservation value.	X	
8	Seek to improve soil health and minimize degradation.	X	
9	Optimize water use, minimize contamination and depletion, and respect water rights.	X	
10	Minimize air pollution.	X	
11	Must be produced cost-effectively.		
12	Do not violate land rights.		X

TABLE 2 Summary of the 25x25 Sustainability Principles and Applicable Levels of the Marshall-Toffel Sustainability Hierarchy

Hierarchy	Principles	Levels 1-3	Level 4
1	Producers and consumers should have equal access to renewable energy markets, products, and infrastructure.		
2	Renewable energy production should maintain or improve air quality.	X	
3	Renewable energy production should maintain or improve biodiversity.	X	
4	Renewable energy production should bolster the local economic foundation and quality of life.		X
5	Renewable energy production should be energy efficient and conserve natural resources.	X	
6	Renewable energy production should reduce greenhouse gas emissions compared to fossil fuels.	X	
7	If invasive species are used, appropriate safeguards should be implemented.	X	
8	Renewable energy production should have market parity with fossil fuels.		
9	All regions of the nation should have the opportunity to participate in renewable energy development and use.		
10	If renewable energy is produced on private land it should improve the health and productivity of these lands.		
11	Renewable energy production on public lands should be sustainable and contribute to the long-term health and mission of the land.		
12	Renewable energy production should incorporate the best available erosion management properties.	X	
13	Renewable energy production should maintain or enhance soil quality.	X	
14	Renewable energy production should respect areas with important conservation, historic, and social value.		X
15	New technologies should be implemented with care to avoid negative consequences.	X	
16	Renewable energy production should maintain or improve water quality.	X	
17	Renewable energy production should maximize water conservation.	X	
18	Renewable energy production should maintain or enhance wildlife habitat health and productivity.	X	

Principles 2, 3, 5, 6, 7, 12, 13, 15, 16, 17, and 18 of the 25x25 action plan all relate directly to either human health or survival, human rights, or species extinction; therefore, they are included in the first three levels of the sustainability hierarchy. Principles 4 and 14 relate to quality of life or values and beliefs; therefore, they fall into the fourth level of the hierarchy. Principles 1, 8, and 9 relate to economic competition and do not fit within the sustainability hierarchy. Principles 10 and 11 state that the some of the principles laid for all lands in other points should also apply to public and private lands. Since a general framework such as this is usually assumed to apply to both public and private lands, restating this in two added principles with slightly differing wording is redundant from the standpoint of sustainability.

United Nations Sustainable Bioenergy Framework for Decision Makers

UN-Energy, a collaborative framework of the United Nations (UN) bodies that contribute to energy solutions, provides a set of principles to draw attention to the "issues that need further attention, analysis, and valuation, so that appropriate trade-offs can be made and both the energy needs of people met and local and global environment adequately protected." The UN framework reports that principles should be created around the issues in Table 3.

Principles 1, 3, 5, 8, and 9 of the UN framework all relate directly to either human health or survival, human rights, or species extinction; therefore, they are included in the first three levels of the Marshall-Toffel Sustainability Hierarchy. Principles 2, 4, 6, and 7 all relate to the economic situation of developing nations. In some cases, the job and economic status of the citizens of developing nations would have a significant impact on the health and survival of those citizens. In those cases, these principles would also be included within the first three levels of the hierarchy. Otherwise they would fall under level four.

Criteria and Indicators

Multiple criteria and indicators are available for biofuels, although the publications that directly apply to the United States are still in draft form. The Inter-

TABLE 3 Summary of the United Nations Sustainable Bioenergy Framework and Applicable Levels of the Marshall-Toffel Sustainability Hierarchy

Hierarchy	Issues	Levels 1-3	Level 4
1	Ability of modern bioenergy to provide energy services to the poor.	X	
2	Implications for agro-industrial development and job creation.	?	X
3	Health and gender implications of modern bioenergy.	X	
4	Implications for the structure of agriculture.	?	X
5	Implications for food security.	X	
6	Implications for government budget.	?	X
7	Implications for trade, foreign exchange balances, and energy security.	?	X
8	Impacts on biodiversity and natural resource management.	X	
9	Implications for climate change.	X	

American Development Bank developed a Biofuels Sustainability Scorecard.[1] The two draft sets of criteria and indicators being developed in the United States are the California Low-Carbon Fuel Standard and the U.S. Renewable Fuel Standard (RFS). The Council on Sustainable Biomass Production also has a draft standard that focuses on dedicated fuel crops, crop residues, purpose-grown wood, and forestry residues in North America. And the Global BioEnergy Partnership is developing a set of sustainability criteria and indicators.

Delzeit and Holm-Müller (2009) published a general guide to developing criteria and indicators, which evolved from their work in developing criteria and indicators for Brazilian bioethanol certification. They found that, in general: (1) sustainability criteria should be grounded in theory, important to stakeholders, and verifiable at a reasonable cost; (2) some criteria that are highly important cannot be included as indicators because of low verifiability; (3) it is difficult to develop a reliable indicator for greenhouse gas reduction; and (4) "Land Conversion Burden" multipliers can be assigned to account for land-use change.

European Frameworks

Some European sustainability frameworks that are complete with principles, criteria, and indicators are the European Union's Biofuel Directive and Fuel Quality Directive (European Commission, 2008), the United Kingdom's Renewable Transport Fuel Obligation (RTFO, 2007), the Netherlands' framework (Cramer et al., 2006, 2007) and World Wildlife Fund Germany (Fritsche et al., 2006). These frameworks are not discussed in depth here.

Certification Schemes

Certification schemes have been developed to solve what is called the Principal-Agent Problem: where potential consumers of biomass (principals) have little or no information about the production characteristics of the products they buy, although those characteristics may be important to the consumer. In the case of biofuels, certification of a brand allows customers to know that the fuel they are buying was produced with a certain amount of sustainability. Delzeit and Holm-Müller (2009) give an example of the certification process for Brazilian bioethanol. Some limitations of certification are potential conflicts with World Trade Organization rules and free trade agreements and the possibility that certification schemes may be used as nontariff trade barriers, requiring standards of practice for production that developing countries do not have the resources to adhere to.

[1] For more information, see *http://www.iadb.org/scorecard*.

TOOLS FOR THE QUANTIFYING ECONOMIC AND ENVIRONMENTAL IMPACTS

The preceding sections have provided sustainability definitions and principles. Although the related criteria and indicators are not discussed in detail in this paper, determining whether a particular biofuel production pathway meets sustainability criteria typically requires quantitative analysis. The software models discussed in this section are among those that can be used for this analysis. Here, the models are divided into greenhouse gas and non-greenhouse gas software.

Greenhouse Gas Software

Life-cycle analysis (LCA) is the process of examining a product or good from cradle to grave, accounting for all of the inputs and outputs during the production, use, and disposal of the product. In the context of transportation fuels, LCA is primarily focused on accounting for the processes that are involved in resource production, refining, transportation, storage, and use of the fuels.

Initially, life-cycle analyses focused on determining the total amount of energy required to produce a fuel. As concerns of climate change have increased and the life-cycle energy of fuels was better understood, much of the attention has shifted to focus on greenhouse gas emissions from fuels. Within the last several decades, numerous software packages have been developed that contain databases of relevant fuel life-cycle data and frameworks for accounting for greenhouse gas emissions. The following sections outline several software packages that calculate fuel life-cycle emissions and/or use LCA emissions as a part of their framework.

GREET

The greenhouse gases, regulated emissions, and energy use in transportation (GREET) model was developed by Argonne National Laboratories to calculate the full life-cycle emissions and energy use from the transportation sector. The model is among the most reviewed of the U.S. models and has been used in many peer-reviewed studies (Farrell et al., 2006; Farrell and Sperling, 2007; Wang et al., 2007; Hill et al., 2009). The model is composed of two separate spreadsheet-based modules that calculate the emissions associated with the well-to-wheels production of fuels (current model 1.8C) and the vehicle production and disposal cycle (current model 2.7).

GREET calculates the energy consumption (with delineated fossil fuel and petroleum consumption) for the entire fuel life cycle. The current model contains more than 100 distinct fuel life cycles. For each fuel pathway the model calculates emissions of five criteria pollutants (volatile organic compounds, carbon monoxide, nitrogen oxides, particulate matter with diameters of 10 micrometers or less, and sulfur oxides) and three greenhouse gases (CO_2, methane [CH_4], and nitrous oxide [N_2O]), along with to the total energy consumption. The GREET model converts all greenhouse gas emissions to carbon dioxide equivalent (CO_2e) emissions based on the Intergovernmental Panel on Climate Change's (IPCC's) global warming potential, which normalizes the radiating forcing of the gases over a 100-year period. In addition to calculating life-cycle emissions of fuels, GREET calculates life-cycle emissions associated with the production and disposal of six different vehicle configurations based on the same vehicle platform. Typical results indicated that emissions from vehicle production and disposal make up ~10 percent of total vehicle use emissions (Wang, 1999).

While GREET accurately accounts for the direct emissions from fuel production, some of the model's indirect emissions calculations need further work. GREET likely underestimates the emissions that result from direct land-use changes and does not calculate any emissions from indirect land-use changes (Farrell and Sperling, 2007). These indirect land use emissions have been shown to be a significant portion of the fuel's lifecycle and may ultimately determine whether ethanol has lower net emissions than gasoline (Fargione et al., 2008; Searchinger et al., 2008). Additionally, co-products created as a part of the fuel production are not well accounted for within GREET. GREET accounts for co-product credits by having set displacement coefficients. It likely overestimates the amount of credits because the true amount of goods that are displaced is a result of market forces, which can only be captured by economic modeling (Farrell and Sperling, 2007).

To deal with the uncertainty of input data, GREET includes a stochastic modeling package that defines probability distributions of critical inputs. Unfortunately, as GREET inputs come from government data, academic literature, and stakeholder input, the probability distributions of the data are rarely known. While using unknown probability distributions to calculate confidence intervals for fuels' emissions can lead to a false confidence in the results, the stochastic modeling package is useful for determining critical parameters that affect emissions. In addition to stochastic modeling, GREET includes a time series feature that allows for projections of future energy use and emissions for the production of fuels. However, results are highly speculative as they are largely influenced by assumptions about the future.[2]

EBAMM

The Energy Resources Group Biofuel Analysis Meta-Model (EBAMM) was developed by the University of California (UC), Berkeley as a model for comparing LCA software. EBAMM was originally developed to compare studies of total energy use of corn ethanol production. Though it only contains inputs for the production of gasoline, corn ethanol, and cellulosic ethanol, the model con-

[2]For more information, refer to *http://www.transportation.anl.gov/modeling_simulation/GREET/*.

sists of inputs from a range of sources and therefore produces a range of values for energy consumption and greenhouse gas emissions. EBAMM does not track non-greenhouse gas pollutants, such as particulate emissions or volatile organic compounds.

EBAMM takes as inputs the results of six previous ethanol studies and compares them for a common set of boundary conditions and assumptions. Results from a study using EBAMM indicate that ethanol requires much less petroleum than gasoline to produce but is nearly equivalent in terms of greenhouse gas emissions (Farrell et al., 2006). A second study using EBAMM showed that biomass was better used to reduce total greenhouse gas emissions by displacing coal in co-fired burners to generate electricity than by displacing gasoline by producing ethanol. Results also indicated that electricity could provide more vehicle miles per hectare when converted to electricity than when converted to ethanol (Campbell et al., 2009).[3]

Peek/Poke and MOUSE

Lifecycle Associates, a private environmental consulting firm, has created two add-on packages for the GREET model. Peek/Poke and Matrix Organization Using Specific Energy (MOUSE) software packages work with GREET to produce or process company-specific fuel life cycles, rather than industrial averages. Peek/Poke serves as a driver for GREET, allowing the user to introduce input data into the software and run simulations without having to modify the GREET code directly. The model first "pokes" the user-defined inputs into the GREET model via Visual Basic macros. Then the software runs the GREET simulation and "peeks" at the results by outputting them from the GREET report.

The MOUSE software works with GREET results to provide accurate accounting of mixed fuels that are not contained within GREET. MOUSE contains a matrix of GREET-calculated fuel life-cycle emissions and allows users to determine emissions for mixtures of fuel types, such as E85 (85 percent ethanol in diesel fuel). The software is designed to help blenders and fuel producers calculate emissions of fuel mixtures that are specific to their processes, compositions, and regions.[4]

BEACCON

The Biofuels Emissions and Cost Connection (BEACCON) model was developed by Richard Plevin at UC Berkeley to calculate the costs of greenhouse gas reductions from ethanol. To create an economic cost model for ethanol production, BEACCON combines the operating and maintenance costs of corn

[3] For more information, refer to *http://rael.berkeley.edu/EBAMM/*.

[4] For more information, refer to *http://www.lifecycleassociates.com/4.html*.

farming and ethanol refining with corresponding emissions from GREET into a single spreadsheet. By combining ethanol production costs with emissions, BEACCON allows users to model the effect of carbon pricing policies on ethanol prices. BEACCON has been used to model the change in the price of ethanol as a result of a charge per unit of life-cycle CO_2 emissions, a charge per unit of direct biorefinery emissions only, and a low-carbon fuel standard (Plevin and Mueller, 2008). Results from the study indicated that costs largely depended on the refinery fuel choice, with natural gas plants incurring low or negative additional costs and coal plants incurring higher costs. Currently the model only contains economic data for ethanol; further development would be required to expand the analysis to other fuels.[5]

LEM

The Lifecycle Emissions Model (LEM) was developed by Mark Delucchi at UC Davis. While the LEM has been used for a variety of studies and is the basis for the GHGenius model (see below) the model itself has not been published in any peer-reviewed journals, so its validity remains unverified. The LEM is not publicly available for independent use, but the model results along with critical inputs are available in a series of reports (Delucchi, 2003, 2004, 2005).

The LEM calculates emissions for fuels from the largest numbers of countries of the LCA models and includes inputs for 30 different countries. The available data for different countries vary in accuracy and completeness, and the model is most complete for use within the United States (Delucchi, 2003). The model is spreadsheet based and currently calculates emissions for 28 fuel pathways and over 20 different vehicles, including passenger vehicles, buses, scooters, bicycles, heavy rail, light rail, diesel trains, and cargo ships. The LEM calculates emissions for 12 pollutants, more than any other LCA model. It contains historical data that allow for results to be calculated for any target year from 1970 to 2050. The historical data also allow the LEM to make predictions about future fuels based on historical data using the model's dynamic capabilities (Delucchi, 2003).

Of all the models, the LEM has the most complete treatment of land-use change. Results from the LEM indicate that the largest sources of cultivation and land-use emissions are: changes in soil carbon and biomass carbon due to cultivation; changes in soil and biomass carbon due to fertilization of off-site ecosystems by all nitrogen input; N_2O emissions from fertilizer use, crop-residues, and biological fixation; and emissions of oxides of nitrogen (Farrell and Sperling, 2007). As a result of more complete inclusion of land-use emissions, the LEM-

[5] For more information, refer to *http://plevin.berkeley.edu/biofuels/*.

calculated emissions are higher than GREET for fuels derived from biomass, but lower than conventional fuels.[6]

GHGenius

GHGenius is an LCA model that focuses on calculating fuel emissions from Canada, although it also contains data specific to the United States and Mexico. The GHGenius model is based on an early version of the LEM, which was modified to include Canadian-specific data in 1998. GHGenius is publicly available and is primarily used by researchers within Canada for the calculation of fuel life-cycle emissions. GHGenius focuses on the calculation of past, current, and future fuel emissions using historical trends as a guide for future emissions. For U.S.-produced fuels, GHGenius shares many of the data sources of the GREET model, such as data from the U.S. Energy Information Administration. For data not tracked by government agencies, GHGenius relies first on industry average values, then on actual operating plant data, engineering design data with simulations, and finally scientific experimentation data as a last resort. As GHGenius does not rely on stakeholder input to the extent GREET does, the calculated emissions for certain fuels deviate largely between the two models.

Like GREET, GHGenius allows for easy manipulation of the input data, so that emissions can be calculated for user-specified fuels. GHGenius's default inputs contain regionally distinct inputs throughout the United States, Canada, and Mexico to discern geographic differences in the biofuel emissions. GHGenius tracks emissions for the same five regulated emissions and three greenhouse gases as GREET, but includes two other greenhouse gases, chlorofluorocarbons (CFC-12) and hydrofluorocarbons (HFC-134a). Currently GHGenius contains a limited number of 28 fuel pathways, when compared to other large models. Of all the models, GHGenius contains the most detailed examination of gasoline and diesel produced from Canadian oil sands.

GHGenius contains a more detailed accounting than GREET does of emissions that result from land-use changes. To do so, GHGenius tracks changes in soil carbon content due to cultivation, fertilizer application, and changes in biomass levels. The largest contributions to GHG emissions from changes in land use are those that result from the use of nitrogen-based fertilizers. Like other LCA programs, GHGenius provides the option to output all GHG emissions in terms of CO_2-equivalent emissions in accordance with the IPCC's 100-year global warming potentials.

Results from GHGenius tend to be very similar to results from the LEM. When compared to the GREET model, emissions from traditional fuels tend to be lower and emissions from biomass-derived fuels tend to be higher. Unlike the

[6]For more information, refer to *http://www.its.ucdavis.edu/publications/2002/UCD-ITS-RR-02-02.pdf.*

LEM and GREET, GHGenius assumes that there are not significant energy efficiency gains at the corn ethanol refinery plants, and the levels of GHG emissions produced in farming grass-based crops are similar to those used to produce corn grain (Warner, 2009). GHGenius contains a Monte Carlo simulation package for the analysis of uncertainty that is comparable to GREET's.

GHGenius also includes a useful function that allows for cost comparison of fuels based on user inputs of fuel and vehicle prices. Results from the output allow the user to evaluate emission reductions per dollar, which is useful for planning and policy decisions.[7]

BESS

Biofuel energy system simulator (BESS) is an LCA model developed by the University of Nebraska. The model specializes in calculating the well-to-tank emissions of corn ethanol produced in different U.S. states and regions. Like other models, BESS is spreadsheet based, but incorporates a user-friendly interface that allows for easy updating of data inputs.

BESS tracks three greenhouse gases: CO_2, N_2O, and CH_4. Like other LCAs, BESS combines emissions into CO_2-equivalent emissions using the IPCC's 100-year global warming potential. In addition to greenhouse gases, BESS tracks the use of other resources necessary for the production of ethanol, including water, which other models neglect to track. The model calculates emissions and resource use from four separate modules that consist of crop production, ethanol biorefining, cattle feeding, and anaerobic digestion (Liska et al., 2008). The inclusion of anaerobic digestion is unique to BESS and highlights the importance of accounting for co-product credits within the life cycle of fuels.

Like other models, BESS relies on government data for agricultural and biorefinery inputs. BESS deviates from other models by including recent state and regional biorefinery efficiencies for the Midwest, which are more efficient than nationwide refinery efficiencies. BESS uses results from other models, primarily GREET, for inputs not within its limited scope, such as emissions from electricity production.

Results from BESS using model defaults indicate substantially lower emissions for corn ethanol when compared to other models (Liska et al., 2009). However, a recent study comparing the inputs for both BESS and GREET shows that BESS undercounts emissions in several key areas, including electricity emissions, fossil fuel factors, and lime fertilization rates. After properly accounting for the omissions within BESS, Plevin (2009) showed that results from BESS are indeed close to the emissions calculated by GREET, when GREET is modified to model natural gas biorefineries only. While BESS provides a positive step forward for

[7]For more information, refer to *http://www.ghgenius.ca/*.

calculating region- and plant-specific emission, the model requires further refinement before widespread use is warranted.[8]

CONCAWE

The joint European study of life-cycle fuel emissions was conducted by the European Council for Automotive R&D, the European Commission Joint Research Centre, and Conservation of Clean Air and Water in Europe (CONCAWE). The study includes over 40 liquid fuels and multiple electricity production pathways, all of which focus on fuels as derived for the European market. The CONCAWE model is not publicly available, but results and inputs are detailed in a series of online reports (Armstrong et al., 2002).

The model builds on previous studies dating back to 1995 and includes updated inputs primarily from government agencies and academic literature (Armstrong et al., 2002). The study includes fuel pathways not seen in models focusing on North American fuels, particularly those derived from rapeseed, sugar beets, and wheat. The model is distinct from American models, in that it does not include significant portions of crude oil sources from Canadian oil sands, as it projects that Europe will continue to use sweet crude through 2030. Although the CONCAWE model does not include a detailed account of emissions as a result of direct or indirect land use-changes, creators of the model acknowledge that land-use changes are significant (Farrell and Sperling, 2007).

Results from the model indicate that emissions from conventional petroleum products have slightly lower emissions in Europe than the United States, however it is not clear whether this indicates a true difference in emissions or is a result of variations in the models. Comparison of CONCAWE biofuels emissions to the emissions of the American models is difficult since the CONCAWE model does not provide corn ethanol pathways and American models do not typically include European type biofuels. The model is currently being updated to incorporate new data and preliminary results have been posted on the CONCAWE website (see links below).[9]

U.S. EPA

The U.S. Environmental Protection Agency (EPA) recently proposed changes to the RFS, which included guidelines that establish standards for evaluating fuel life-cycle greenhouse gas emissions. As a part of the RFS framework, qualifying biofuels are required to reduce life-cycle emissions by defined percentages relative to the traditional fossil fuels that they replace. Life-cycle emissions for both traditional and biomass-derived fuels will include all direct and indirect emis-

[8]For more information, refer to *http://www.bess.unl.edu/*.

[9]For more information, refer to *www.concawe.org* and *http://ies.jrc.ec.europa.eu/WTW*.

sions, including land-use changes. To determine the full life cycle of fuels, EPA is using a combination of fuel, agricultural, and economic models.

The EPA guidelines indicate that the new standard used a combination of models to determine life-cycle emissions, including GREET, Texas A&M University's Forestry and Agricultural Sector Optimization Model, Iowa State University's Food and Agricultural Policy Research Institute's (FAPRI's) international agricultural models, and the Winrock International database (U.S. EPA, 2009).

The rulemaking process is still ongoing, and EPA is seeking input for the rule. Despite the uncertain nature of the final rule, the EPA has indicated several important aspects of the framework that the final rule will contain. For example, EPA has decided that the overall fuel life cycle will include greenhouse gas emissions released both domestically and internationally as a result of U.S. fuel consumption. The full life-cycle emissions from fuels are evaluated according to their incremental increase in production volume to comply with the 2022 RFS requirements, rather than focusing on a specific gallon of fuel. EPA's analysis does not distinguish emissions within a given feedstock—i.e. all corn production will have the same average value of emissions associated with the life cycle, regardless of where and how it is grown. EPA also states that the uncertainty of aspects of the fuel life cycle does not warrant their exclusion from the model—i.e. international land use and nitrogen cycles (U.S. EPA, 2009). Significant controversy remains about how land-use changes will be incorporated in the final rule and what time frame for evaluating payback horizon will be used to evaluate land-use changes (Grunwald, 2009).[10]

Swiss Life-Cycle Assessment of Energy Products

The Swiss government developed a method for evaluating fuels to determine the full energy, greenhouse gas, and environmental costs of transportation fuels used in Switzerland, The method uses a life-cycle assessment model based on input data from Ecoinvent 1.3 to determine the overall energy consumption and greenhouse gas emissions of the fuels. The method also evaluates fuels on their overall environmental impacts using Eco-Indicator 99 and Environmental Impact Points, UBP 06. The Swiss method also reports the impact of fuels with two metrics, greenhouse gas emissions and overall environmental impacts, which include damage to human health, damage to ecosystems, and depletion of nonrenewable resources (Zah et al., 2007).

Findings from the Zah et al. study indicate that while most fuels derived from biomass have reduced emissions when compared to petroleum-derived fuels, they often have overall environmental impacts that are significantly more severe (Zah, 2007). In particular, the study highlights how fuels produced from crop mono-

[10]For more information, refer to *http://www.epa.gov/otaq/renewablefuels/index.htm*.

cultures (such as corn ethanol) have substantially higher environmental costs in terms of eutrophication, acidification, and land occupation and transformation. Limitations of the Swiss method stem from the use of an old dataset (2004) and combination of dissimilar environmental impacts into a single metric that reflects subjective rather than universal environmental values.[11]

LEAP

The Long-range Energy Alternatives Planning System (LEAP) is a software program developed by the Stockholm Environmental Institute for conducting long-range energy and emissions planning. LEAP is not an LCA model of fuel production. Rather, it provides a framework for analyzing fuel emissions as they relate to vehicle efficiency and use. LEAP provides baseline tailpipe emissions for different fuels that can easily be augmented to include total fuel life-cycle emissions when used in conjunction with fuel LCA software.

LEAP incorporates fuel emissions with other key components of the transportation sector to solve for regional, state, or national emissions from the transportation sector. By including such components as vehicle mix and turnover rates, it is possible to determine how fast specific fuels, such as E85, can penetrate the market. A recent analysis using the LEAP framework within Minnesota examined how separate policies aimed at reducing vehicle fuel consumption, life-cycle fuel emissions, and vehicle miles traveled combine to reduce overall emissions (Boies et al., 2008). In addition to tracking GHG emissions, LEAP accounts for regulated pollutants and can be modified to accounts for other critical factors, such as water use.[12]

Energy Choice Simulator

The Energy Choice Simulator was developed by the Great Plains Institute and the University of Minnesota to model the effect of various fuel policies on the price, quantity, and emissions from the transportation fuels sector. The Energy Choice Simulator is a Web-based tool that draws on outputs from GREET to calculate the change in full life-cycle greenhouse gas emissions based on changes to future policies. In addition to greenhouse gas emissions, the simulator tracks the same regulated pollutants that are included in GREET.

The Energy Choice Simulator currently includes data for 12 states in the Midwest, including life-cycle fuel emissions, information on vehicle fleet makeup, vehicle turnover rate, and current and proposed policies. The Web-based model allows users to test assumptions about future policies and compare them to a base case scenario. Policies that are included are individual or regional state taxes

---

[11]For more information, refer to *http://www.bioenergywiki.net/images/8/80/Empa_Bioenergie_ExecSumm_engl.pdf. http://www.sciencemag.org/cgi/content/short/319/5859/43.*

[12]For more information, refer to *www.energycommunity.org.*

and subsidies, low-carbon fuel standards, and efficiency mandates. The simulator is currently under development and is expected to be available in late 2009. (Warner, 2009).[13]

Non-GHG Software

Agricultural Models

POLYSYS—The Policy Analysis System (POLYSYS) is an economic simulation modeling system of the U.S. agricultural sector. POLYSYS incorporates agricultural planning decisions in each of 305 U.S. agricultural statistical districts and national averages for crop demands and prices as well as livestock supply and demand. Using the agricultural data, POLYSYS estimates agricultural production response, resource use, price, income, and environmental impacts of projected changes from an agricultural baseline. POLYSYS is able to model the first- and second-generation biofuel crops of corn, soybeans, sugarcane, switchgrass, hybrid poplars, and hybrid willows, among other crops. POLYSYS is a partial-equilibrium model, meaning that it considers the interrelatedness of the agricultural sector with some other sectors, but not all sectors. For instance, it can model the interdependence of biofuel production and prices with livestock feed production and prices, but it cannot model the interdependence of biofuel production and prices with oil production prices.

POLYSYS uses a baseline approach, meaning that it simulates a deviated path from a published agricultural baseline. This approach allows for quick turnaround and relatively few data requirements, as the majority of the simulation work has already been done to generate the baseline. POLYSYS baseline data are available from the U.S. Department of Agriculture (USDA), FAPRI, and the U.S. Congressional Budget Office. It can be coupled with IMPLAN (Impact Analysis for Planning) to model income data and with EPIC (Environmental Policy Integrated Climate) to model environmental impacts (both discussed below). If EPIC is used, POLYSYS uses data from the USDA STATSGO and GRASS databases and selection criteria from the USDA Natural Resources Conservation Service to identify dominant soils. For simulations involving crop rotations and cropping practices, data can be obtained from the USDA Cropping Practices Survey, and for simulations involving enterprise or rotation budgets, data can be obtained from the Agricultural Policy Analysis Center Budgeting System.

POLYSYS has been used to estimate the potential U.S. biomass feedstock supply (Ugarte et al., 2000; Walsh et al., 2003) and the economic and agricultural sector impacts of a potential increase in demand for biodiesel (Ugarte et al., 1999). It has also been used along with REAP (discussed below) to quantify the

[13]For more information, refer to *http://forio.com/simulation/mga/index* and *http://forio.com/wiki/mga/index.php/Main_Page*.

environmental and economic impacts of increased U.S. biofuel feedstock production (BRDI 2008). POLYSYS allows a detailed simulation of land-use change effects within the continental United States.

The POLYSYS model has some limitations. Since POLYSYS simulations are anchored to a baseline, the accuracy of all results is dependent on the accuracy of the baseline. POLYSYS is a deterministic, not stochastic, model, and as such is not able to calculate probability distributions of different outcomes. As mentioned above, the model cannot simulate the interdependency of the energy and agricultural sectors. Therefore, it is necessary to model the increase in bioenergy production in the form of a mandate, rather than in reaction to energy prices. POLYSYS also cannot simulate forestland, and thus cannot model biomass production from forest residue. Since POLYSYS only models the continental United States, it cannot simulate international land-use changes.[14]

REAP—The Regional Environment and Agricultural Programming Model (REAP) is a partial-equilibrium agricultural model, similar to POLYSYS. REAP simulates how changes in policy, demand, and production technology affect the regional supply of crops and livestock, commodity prices, crop management behavior and the use of production inputs, farm income, and environmental indicators. Similar to POLYSYS, REAP's results are relative to a baseline projection, and REAP uses EPIC (discussed below) to simulate biophysical indicators.

REAP also shares some limitations with POLYSYS. The accuracy of its results is constrained by the accuracy of the baseline, and REAP cannot calculate stochastic distributions, simulate interdependency between the energy and agricultural sectors, simulate forest land, or model land-use changes outside of the continental United States.

REAP is different from POLYSYS in that it is a static framework: it assesses changes in market and other conditions for a given point in time. While POLYSYS can show the impact path of a certain event over time relative to a baseline, REAP can only calculate a snapshot equilibrium. REAP also only has information for first-generation biofuel crops of corn, soybeans, and sugarcane. It is not able to model second-generation biofuel crops.

The data requirements for the 50 U.S. agricultural regions modeled by REAP are crop yields, input requirements, costs, and returns. The data are provided by USDA's Agricultural Resource Management Survey and the EPIC model. REAP has been used along with POLYSYS by the Biomass Research and Development Initiative (BRDI, 2008).[15]

---

[14]For more information, refer to Ugarte and Ray (2000) and *http://www.agpolicy.org/polysys.html*.

[15]For more information refer to BRDI (2008) and *http://www.ers.usda.gov/Publications/TB1916/*.

Environmental Models

EPIC—The Environmental Policy Integrated Climate (EPIC) model was developed by USDA to simulate the impact of agricultural management strategies on agricultural production and soil and water resources. EPIC takes soil, weather, and management information as inputs and outputs crop yields, erosion, and chemical discharges to the environment. EPIC is used within POLYSYS and REAP to calculate biophysical properties on a field-by-field basis.

The version of EPIC included in REAP and POLYSYS is limited by its use of historical weather data. Farm chemical runoff and erosion occur disproportionately during extreme weather events. Since climate instability and the frequency of extreme weather events are projected to increase in the future, the use of historical weather data may decrease the accuracy of runoff and erosion predictions (BRDI, 2008). However, the standalone version of EPIC has been used to simulate the effects of global climate change on crops (Gassman et al., 2005).

EPIC is only able to simulate the properties of a single field. Therefore, to be able to model the impact of an agricultural simulation on a watershed, EPIC would need to be linked to SPARROW (described below). This is yet to be accomplished.

SPARROW—The U.S. Geological Survey's (USGS's) SPAtially Referenced Regression On Watershed Attributes (SPARROW) model links water quality with constituent sources. SPARROW uses USGS land-use and land-cover data and USDA data on animal nutrients and cropland area. It tracks the transport of nitrogen from atmospheric deposition, nitrogen and phosphate from agricultural fertilizer, and nutrients from urban and other runoff as they are transported to streams and downstream receiving waters. It also tracks the attenuation of these nutrients by natural processes as they are transported from land and downstream.

SPARROW has stochastic capabilities to predict the uncertainty embedded in its simulations. SPARROW can predict water quality in both small watersheds and large river drainages. From a policy standpoint, SPARROW can be used to predict the changes in water quality due to management actions or changes in land use.

SPARROW also has limitations. Due to data limitations, it cannot account for effects of land management or conservation practices, manure application, or urban contaminants (i.e., sewer overflows). SPARROW's mean load spatial distributions are disproportionately influenced by high-flow data; therefore, the mean spatial distributions are more indicative of high-flow seasons than of other times of the year. SPARROW predicts long-term averages, not short-term values, and is more accurate across broad regions than in single catchments.[16]

---

[16]For more information, refer to *http://water.usgs.gov/nawqa/sparrow/*.

Economic Models

RIMS II—The Bureau of Economic Analysis (BEA) Regional Input-Output Modeling System (RIMS II) is a tool for estimating the indirect impacts of changes in a local economy. RIMS II acts as a multiplier: users provide the initial effects in output, earnings, or employment of a change, such as closing an army base or opening an ethanol plant, and RIMS calculates the total impact on output, earnings, or employment over a region of specified size that is at least one county. The RIMS multipliers use data from BEA's national input-output table of 500 industries and BEA's regional accounts.

As models discussed so far, the RIMS multipliers have limitations. Although studies have found that RIMS gives similar results to more complex input-output models,[17] RIMS is only recommended for use with small-scale changes. The national-scale ramping up of biofuel production is beyond the scope of this tool. Also, as reported by Swenson (2007) and Low and Isserman (2009), RIMS II does not have an appropriate category for biofuels. They fall under the larger category of "organic chemical industry," which does not have sufficiently similar characteristics to those of biofuel production plants.[18]

IMPLAN—The Impact Analysis and Planning (IMPLAN) model, developed by the Minnesota IMPLAN Group, is also an input-output model, but is more complex than RIMS II. Like RIMS II, IMPLAN models the total regional economic effect of a given change, but IMPLAN splits the additional effects beyond the initial action into two categories: indirect and induced. Indirect effects are changes in interindustry transactions, or basically the supply and distribution chains of the affected entity. Induced effects are the changed spending habits in the local economy. IMPLAN can also disaggregate impacts into sectors of the economy. IMPLAN requires data from the U.S. system of national accounts, which are collected by the U.S. Department of Commerce's Bureau of Labor Statistics and other federal and state government agencies.

Even when using a more sophisticated tool such as IMPLAN, it is possible to misportray the number of local jobs created by increased biofuel production. As explained in Swenson (2007) and Low and Isserman (2009), the corn produced for ethanol is sometimes misclassified as new production, while in reality virtually all of the production is pre-existing and simply diverted from other uses. This misclassification by itself can cause a 200 percent overestimate in the number of jobs created (Swenson, 2007). Also, increased profits from the price premiums given by ethanol plants do not necessarily stay in the hands of the farmers. Rather, if the farmers do not own their land, excess profits go to the landlords in the form of increased rent, many of whom do not reside locally (Low and Isserman, 2009). Also, IMPLAN assumes that soybean production is more

[17] See *http://www.bea.gov/regional/rims/brfdesc.cfm*.

[18] For more information refer to Swenson (2007) and *http://www.bea.gov/regional/rims/brfdesc.cfm*.

labor intensive than corn production, which is not true on a local level. Farmers who decide to plant soybeans in a given year instead of corn typically do not hire extra workers to do so (Low and Isserman, 2009).

Air Quality Models

Examples of models that could be used to estimate the impacts of biofuels on air quality are CAMx,[19] CMAQ,[20] and GATOR-GCMM (Jacobson, 2001). These models incorporate emissions, meteorology, and photochemical reactions. Applications of these models to biofuels include Jacobson (2007), Hill et al. (2009), and Morris et al. (2003)

SUMMARY

This document has reviewed the proposed and extant frameworks to explore and label the sustainability of biofuel production and the software tools available to quantify different aspects of that sustainability. Frameworks of sustainability have been discussed and, while no one definition is universally applicable, the Marshall and Toffel Sustainability Hierarchy was used to evaluate sustainability principles in the context of biofuels. Also discussed were criteria, indicators, and certification schemes for biofuels. The study discussed a variety of tools for determining various aspects of the sustainability of biofuels, including greenhouse gas, environmental, economic, and air quality models. Of the greenhouse gas models, GREET was found to be the most widely used and most comprehensive, while other models, such as GHGenius, were found to have better treatment of land-use effects. The agricultural/economic models POLYSYS and REAP were similar in structure, with POLYSYS having the advantage of being able to calculate the impact path of a decision over time. The economics model IMPLAN was found to be more accurate than RIMS II. Overall, it was found that carefully considering the inputs to any economic model is important in obtaining accurate results. Several studies have modeled the air quality impacts of biofuel production, but work in this area is still preliminary.

A U.S. Department of Energy and USDA report offers a more thorough discussion of research frontiers (U.S. DOE and USDA, 2009).

REFERENCES

Armstrong, A. P., et al. April 2002. *Energy and Greenhouse Gas Balance of Biofuels for Europe—An Update.* Report No. 2/02. Brussels: CONCAWE Ad Hoc Group on Alternative Fuels. Available at *www.concawe.be*.

[19]For more information, refer to *http://www.camx.com/*.

[20]For more information, refer to *http://www.cmascenter.org/*.

Boies, A., et al. June 2008. *Reducing Greenhouse Gas Emissions from Transportation Sources in Minnesota.* CTS 08-10. Minneapolis, MN: University of Minnesota, Center for Transportation Studies. Available at *http://www.cts.umn.edu/Research/Featured/GreenhouseGas/.*

BRDI (Biomass Research and Development Initiative). 2008. *Increasing Feedstock Production for Biofuels: Economic Drivers, Environmental Implications, and the Role of Research.* Biomass Research and Development Board. Available at *http://www1.eere.energy.gov/biomass/publications.html.*

Cramer, J., et al. 2006. *Criteria for Sustainable Biomass Production: Final Report from the Project Group "Sustainable Production of Biomass."* Available at *www.globalproblems-globalsolutions-files.org/unf_website/PDF/criteria_sustainable_biomass_prod.pdf* [Accessed June 6, 2009].

Cramer, J., et al. 2007. *Testing Framework for Sustainable Biomass: Final Report from the Project Group "Sustainable Production of Biomass."* Available at *http://www.lowcvp.org.uk/assets/reports/070427-Cramer-FinalReport_EN.pdf* [Accessed June 6, 2009].

Campbell, J. E., et al. 2009. Greater transportation energy and GHG offsets from bioelectricity than ethanol. *Science* 324(5930):1055-1057. Available at *http://www.sciencemag.org/cgi/content/abstract/1168885.*

Delzeit, R., and K. Holm-Müller. 2009. Steps to discern sustainability criteria for a certification scheme of bioethanol in Brazil: approach and difficulties. *Energy* 34(5):662-668.

Delucchi, M. A. 2002. "Overview of the Lifecycle Emissions Model (LEM)." University of California, Davis. Available at *www.its.ucdavis.edu/publications/2002/UCD-ITS-RR-02-02.pdf.*

Delucchi, M. A. 2003. *A Lifecycle Emissions Model (LEM): Lifecycle Emissions from Transportation Fuels, Motor Vehicles, Transportation Modes, Electricity Use, Heating and Cooking Fuels, and Materials.* Institute of Transportation Studies, University of California, Davis. Available at *http://escholarship.org/uc/item/9vr8s1bb.*

Delucchi, M. A. 2004. *Conceptual and Methodological Issues in Lifecycle Analysis of Transportation Fuels.* Research Report UCD-ITS-RR-03-17. Institute of Transportation Studies, University of California, Davis. Available at *http://pubs.its.ucdavis.edu/publication_detail.php?id=203.*

Delucchi, M. A. 2005. *A Multi-Country Analysis of Lifecycle Emissions from Transportation Fuels and Motor Vehicles.* Research Report UCD-ITS-RR-05-10. University of California, Davis. Available at *http://pubs.its.ucdavis.edu/publication_detail.php?id=52.*

EC (European Commission). 2008. European Parliament legislative resolution of 17 December 2008 on the proposal for a directive of the European Parliament and of the Council amending Directive 98/70/EC as regards the specification of petrol, diesel and gas-oil and introducing a mechanism to monitor and reduce greenhouse gas emissions from the use of road transport fuels and amending Council Directive 1999/32/EC, as regards the specification of fuel used by inland waterway vessels and repealing Directive 93/12/EEC.

U.S. EPA (Environmental Protection Agency). 2009. Regulation of Fuels and Fuel Additives: Modification to Renewable Fuel Standard Program; Proposed Rule, p. 74. Available at *http://www.epa.gov/OMS/renewablefuels/.*

Fargione, J., et al. 2008. Land clearing and the biofuel carbon debt. *Science* 319(5867):1235-1238. Available at *http://www.sciencemag.org/cgi/content/abstract/1152747.*

Farrell, A. E., and D. Sperling. 2007. *A Low-Carbon Fuel Standard for California, Part 1: Technical Analysis.* Research Report UCD-ITS-RR-07-07. Institute of Transportation Studies, University of California, Davis. Available at *http://pubs.its.ucdavis.edu/publication_detail.php?id=1082.*

Farrell, A. E., et al. 2006. Ethanol can contribute to energy and environmental goals. *Science* 311(5760):506-508. Available at *http://www.sciencemag.org/cgi/content/abstract/311/5760/506.*

Fritsche, U. R., et al. 2006. *Sustainability Standards for Bioenergy.* Frankfurt am Main, Germany: WWF Germany. Available at *www.wwf.de/.../fm.../Sustainability_Standards_for_Bioenergy.pdf.*

Gassman, P. W., et al. 2005. *Historical Development and Applications of the EPIC and APEX Models.* Center for Agricultural and Rural Development, Iowa State University. Available at *http://ideas.repec.org/p/ias/cpaper/05-wp397.html.*

Graedel, T. E. and R. J. Klee. 2002. Getting serious about sustainability. *Environmental Science & Technology* 36(4):523-529. Available at *http://pubs.acs.org/doi/abs/10.1021/es0106016.*

Grunwald, M. 2009. Stress-testing biofuels: how the game was rigged. *Time* May 12 Available at *http://www.time.com/time/health/article/0,8599,1897549,00.html* [Accessed June 4, 2009].

Hill, J., et al. 2009. Climate change and health costs of air emissions from biofuels and gasoline. *Proceedings of the National Academy of Sciences of the United States of America* 106(6):2077-2082. Available at *http://www.pnas.org/content/early/2009/02/02/0812835106.abstract.*

Jacobson, M. Z. 2001. GATOR-GCMM: global- through urban-scale air pollution and weather forecast model 1. Model design and treatment of subgrid soil, vegetation, roads, rooftops, water, sea ice, and snow. *Journal of Geophysical Research* 106(D6):5385-5401. Available at *www.stanford.edu/group/efmh/jacobson/GATORGCMM101.pdf.*

Jacobson, M. Z. 2007. Effects of ethanol (E85) versus gasoline vehicles on cancer and mortality in the United States. *Environmental Science and Technology* 41(11):4150-4157. Available at *http://pubs.acs.org/doi/abs/10.1021/es062085v.*

Liska, A. J., et al. 2008. "BESS: Biofuel Energy Systems Simulator. A Model for Life-Cycle Energy & Emissions Analysis of Corn-Ethanol Biofuel Production Systems." Available at *http://www.bess.unl.edu* [Accessed June 2, 2009].

Liska, A. J., et al. 2009. Improvements in life cycle energy efficiency and greenhouse gas emissions of corn-ethanol. *Journal of Industrial Ecology* 13(1):58-74. Available at *http://www3.interscience.wiley.com/journal/121647166/abstract?CRETRY=1&SRETRY=0.*

Low, S. A., and A. M. Isserman. 2009. Ethanol and the local economy: industry trends, location factors, economic impacts, and risks. *Economic Development Quarterly* 23(1):71-88. Available at *http://edq.sagepub.com/cgi/content/abstract/23/1/71.*

Marshall, J. D., and M. W. Toffel. 2005. Framing the elusive concept of sustainability: a sustainability hierarchy. *Environmental Science & Technology* 39(3):673-682. Available at *http://pubs.acs.org/doi/abs/10.1021/es040394k.*

Mulkey, D., and A. W. Hodges. 2004. Using IMPLAN to assess local economic impacts. University of Florida IFAS Extension. Available at *http://edis.ifas.ufl.edu/FE168.*

Morris, R. E., et al. 2003. *Impact of Biodiesel Fuels on Air Quality and Human Health: Summary Report.* NREL/SR-540-33793. Golden, CO: National Renewable Energy Laboratory. Available at *www.nrel.gov/docs/fy03osti/33793.pdf.*

Nattrass, B., and M. Altomare. 1999. *The Natural Step for Business: Wealth, Ecology and the Evolutionary Corporation.* Gabriola Island, British Columbia: New Society Publishers. Available at *http://www.newsociety.com/bookid/3656.*

Plevin, R. 2009. Modeling corn ethanol and climate: a critical comparison of the BESS and GREET models. *Journal of Industrial Ecology* 13(4):495-507. Available at *http://www3.interscience.wiley.com/journal/122466949/abstract.*

Plevin, R. J., and S. Mueller. 2008. The effect of CO_2 regulations on the cost of corn ethanol production. *Environmental Research Letters* 3:024003. Available at *www.iop.org/EJ/article/1748-9326/3/2/.../erl8_2_024003.pdf.*

Rees, W., and M. Wackernagel. 1994. *Our Ecological Footprint: Reducing Human Impact on the Earth.* Gabriola Island, British Columbia: New Society Publishers. Available at *http://www.newsociety.com/bookid/3663.*

RTFO (Renewable Transport Fuel Obligation). 2007. Carbon and Sustainability Reporting Within the Renewable Transport Fuel Obligation: Requirements and Guidance. Draft Government Recommendation to RTFO Administrator, Department for Transport. Available at *http://www.dft.gov.uk/pgr/roads/environment/rtfo/* [Accessed June 1, 2009].

Scharlemann, J. P. W., and W. F. Laurance. 2008. Environmental science: how green are biofuels? *Science* 319(5859):43-44. Available at *http://www.sciencemag.org/cgi/content/short/319/5859/43.*

Searchinger, T., et al. 2008. Use of U.S. croplands for biofuels increases greenhouse gases through emissions from land-use changes. *Science* 19(5867):1238-1240. Available at *http://www.sciencemag.org/cgi/content/abstract/1151861*.

Swenson, D. 2007. "Understanding Biofuels Economic Impact Claims." Department of Economics Staff Report, Iowa State University. Available at *www.econ.iastate.edu/research/webpapers/paper_12790.pdf*.

Ugarte, D., et al. 2000. The Economic Impacts of Bioenergy Crop Production on U.S. Agriculture. Oak Ridge, TN: Oak Ridge National Laboratory. Available at *www.usda.gov/oce/reports/energy/AER816Bi.pdf*.

Ugarte, D. G., and D. E. Ray. 2000. Biomass and bioenergy applications of the POLYSYS modeling framework. *Biomass and Bioenergy* 18(4):291-308.

Ugarte, D. G., et al. 1999. *Assessment of Biodiesel Production Potential in the Southeast: Final Report.* Southeast Regional Biomass Energy Program.

U.S. DOE and USDA (U.S. Department of Energy and U.S. Department of Agriculture. 2009. *Sustainability of Biofuels: Future Research Opportunities.* Report from the October 2008 Workshop. Available at *http://genomicsgtl.energy.gov/biofuels/sustainability/* [Accessed May 14, 2009].

Wackernagel, M., et al. 2002. Tracking the ecological overshoot of the human economy. *Proceedings of the National Academy of Sciences of the United States of America* 99(14):9266-9271. Available at *www.pnas.org/content/99/14/9266.full.pdf*.

Walsh, M., et al. 2003. Bioenergy crop production in the United States: potential quantities, land use changes, and economic impacts on the agricultural sector. *Environmental and Resource Economics* 24(4):313-333. Available at *http://ideas.repec.org/a/kap/enreec/v24y2003i4p313-333.html*.

Wang, M. 1999. *GREET 1.5—Transportation Fuel Cycle Model. Volume 2: Appendices of Data and Results.* Argonne, IL: Argonne National Laboratory, p. 208. Available at *http://www.transportation.anl.gov/modeling_simulation/GREET/publications.html*.

Wang, M., et al. 2007. "Life-cycle energy and greenhouse gas emission impacts of different corn ethanol plant types." *Environmental Research Letters* (2): 024001.

Warner, E. S. 2009. Evaluating Life Cycle Assessment (LCA) Models for Use in a Low Carbon Fuel Standard Policy. M.S. thesis. St. Paul, MN: University of Minnesota. Available at *conservancy.umn.edu/bitstream/50481/1/Warner,%20Ethan.pdf*.

Zah, R., et al. 2007. *Lifecycle Assessment of Energy Products: Environmental Assessment of Biofuels.* Bern, Switzerland: Swiss Federal Office for Energy, Environment and Agriculture.

Appendix F

Selected Bibliography

ANALYTIC TOOLS FOR EXAMINING BIOFUELS AND SUSTAINABILITY

Kammen, Daniel, Alexander Farrell, Richard Plevin, Andrew Jones, Mark Delucchi, and Gregory Nemet. 2008. Energy and Greenhouse Impacts of Biofuels: A Framework for Analysis. Organization for Economic Cooperation and Development, Paris, France.

Kearn, S. and N. McCormick. 2008 Implementing Sustainable Bioenergy Production; A Compilation of Tools and Approaches. IUCN. Gland, Switzerland.

National Advisory Council for Environmental Policy and Technology (EPA). 2007. Strategic Framework for Biofuels Efforts.

Roundtable on Sustainable Biofuels, Ecole Polytechnique Federale de Lausanne. 2008. Version Zero—Principles and Criteria. *http://cgse.epfl.ch/webdav/site/cgse/users/171495/public/RSB-brochure-eng.pdf*

Van Dam, Jinke, Martin Junginger, Andre Faaij, Ingmar Jurgens, Gustavo Best, and Uwe Fritsche. 2008. Overview of recent developments in sustainable biomass certification. *Biomass and Bioenergy* 32:749-780.

LIFE CYCLE ASSESSMENTS

Liska, Adam, Haishun Yang, Virgil Bremer, Terry Klopfenstein, Daniel Walters, Galen Erickson, and Kenneth Cassman. 2009. Improvements in Life Cycle Energy Efficiency and Greenhouse Gas Emissions of Corn-Ethanol. *Journal of Industrial Ecology* published online on January 21, 2009. *http://www3.interscience.wiley.com/cgi-bin/fulltext/121647166/PDFSTART.*

Wang, Michael, May Wu, and Hong Huo. 2007. Life-Cycle Energy and Greenhouse Gas Emissions of Different Corn Ethanol Plant Types. *Environmental Research Letters* 2: 024001 (May 22, 2007). *http://www.iop.org/EJ/article/1748-9326/2/2/024001/erl7_2_024001.pdf?request-id=c4950a50-935f-4a1b-8cf9-5191c27271c0.*

Winrock International. 2009. The Impact of Expanding Biofuel Production on GHG Emissions, White Paper #1: Accessing and Interpreting Existing Tools. Arlington, VA.

FEEDSTOCKS AND TECHNOLOGIES

Ebert, Jessica 2008. Breakthroughs in Green Gasoline Production, Biomass Magazine.

Evans, Alexander and Robert Perschel. 2009. An Assessment of Biomass Harvesting Guidelines. Forest Guild. *http://www.forestguild.org/publications/research/2009/biomass_guidelines.pdf.*

MacLean, Heather and Sabrina Spatari. 2009. The Contribution of Enzymes and Process Chemicals to the Life Cycle of Ethanol. *Environmental Research Letters* 4: 014001.

Milbrandt, A. 2005. A Geographic Perspective on the Current Biomass Resource Availability in the United States. National Renewable Energy Laboratory. Technical Report, NREL/TP-560-39181. Golden, CO.

Perlack, Robert D., Lynn L. Wright, and Anthony F. Turhollow. 2005. A Bioenergy and Bioproducts Industry: The Technical Feasibility of a Billion Ton Annual Supply. A report prepared for the United States Department of Energy and the United States Department of Agriculture. Oak Ridge National Laboratory. Oak Ridge, TN. *www.ornl.gov/~webworks/cppr/y2001/rpt/123021.pdf.*

Sandia Laboratories and General Motors Corporation. 2009. 90-Billion Gallon Biofuel Deployment Study. Livermore, CA. (executive summary found at *http://hitectransportation.org/news/2009/Exec_Summary02-2009.pdf.*

ECONOMICS

Babcock, Bruce A. 2009. Intricacies of Meeting the Renewable Fuels Standard, Iowa Ag Review, 4-6.

Eidman, Vernon R. 2007. The Evolving Ethanol Industry in the United States. Paper was prepared for presentation at a session titled "The Economic Impact of Ethanol Development in North America." Canadian Agricultural Economics Society, Portland, OL.

Food and Agriculture Organization of the United Nations. 2009. The State of Food and Agriculture 2008: Biofuels: Prospects, Risks and Opportunities. Rome, Italy.

Fortenberry, T. Randall and Steve Deller. 2008. Understanding Community Impacts: A Tool for Evaluating Economic Impacts from Local Bio-Fuels Production. *Journal of Extension* 46:6. *http://www3.interscience.wiley.com/cgi-bin/fulltext/121496622/HTMLSTART.*

Fortenberry, T. Randall and Hwanil Park 2009. Is Ethanol of Blame for High Food Prices? Status of Wisconsin Agriculture—Special Article: Bioenergy and Agriculture in Wisconsin.

Kleinschmit, Jim. 2007. Biofueling Rural Development: Making for Case for Linking Biofuel Production to Rural Revitalization. Carsey Institute, University of New Hampshire, *Policy Brief* Number 5, Winter 2007.

Leistritz, F. Larry and Nancy M. Hodur. 2008. Biofuels: A Major Rural Economic Development Opportunity. Wiley InterScience (on line) DOI: 10.1002/bbb.104: *Biofuels, Bioproducts and Biorefining.* 2: 501-504. *http://www3.interscience.wiley.com/cgi-bin/fulltext/121426505/PDFSTART.*

Low, Sarah A. and Andrew M. Isserman. 2009. Ethanol and the Local Economy: Industry Trends, Location Factors, Economic Impacts, and Risks. *Economic Development Quarterly* 23:1: 71-88.

Sexton, Steven, Deepak Rajagopal, David Zilberman and Gal Hochman 2008. Food versus Fuel: How Biofuels Make Food More Costly and Gasoline Cheaper. Agricultural and Resource Economics Update, University of California 12:1.

Swenson, David. 2007. Understanding Biofuels Economic Impact Claims. Department of Economics, Iowa State University. Ames, IA. *www.econ.iastate.edu/research/webpapers/paper_12790.pdf.*

Swenson, David. 2008. The Economic Impact of Ethanol Production in Iowa. Department of Economics, Iowa State University. Ames, IA.

Tiffany, Douglas C. 2009. Economic and Environmental Impacts of U.S. Corn Ethanol Production and Use. Federal Reserve Bank of St. Louis. *Regional Economic Development:* 5:1: 42-58.

Tokgoz, Simla, Amani Elobeid, Jacinto Fabiosa, Dermot J. Hayes, Bruce A. Babcock, Tun-Hsiang (Edward) Yu, Fengxia Dong, and Chad E. Hart. 2008. Bottlenecks, Drought, and Oil Price Spikes: Impact on U.S. Ethanol and Agricultural Sectors. *Review of Agricultural Economics* 30:4: 604-622. *www.econ.iastate.edu/research/webpapers/paper_12865.pdf.*

Urbanchuk, John M. 2009. Contribution of the Ethanol Industry to the Economy of the United States. Prepared for the Renewable Fuels Association.

Ugarte, Daniel G. De La Torre, Burton C. English and Kim Jensen 2007. Sixty Billion Gallons by 2030: Economic and Agricultural Impacts of Ethanol and Biodiesel Expansion. American Journal of Agricultural Economics 89:5:1290-1295.

POLICIES AND SUBSIDIES

Doornbosch R. and R. Steenblik. 2007. Biofuels: Is the Cure Worse than the Disease? Roundtable on Sustainable Development. Organization for Economic Cooperation and Development. Paris, France.

Koplow, Doug. 2007. Biofuels—At What Cost? Government support for ethanol and biodiesel in the United States: 2007 Update. Earth Track, Inc. The Global Subsidies Initiative (GSI) of the International Institute for Sustainable Development (IISD)

Office of the Legislative Auditor, State of Minnesota. 2009. Evaluation Report: Biofuel Policies and Programs.

Organization for Economic Cooperation and Development. 2008. Economic Assessment of Biofuel Support Policies. Paris, France. *http://www.oecd.org/document/28/0,3343,fr_2649_33717_41013916_1_1_1_1,00.html.*

Pinchot Institute for Conservation and the Heinz Center. 2009. Ensuring Forest Sustainability in the Development of Wood Biofuels and Bioenergy: Implications for Federal and State Policies. Summary Report. *http://pinchot.org/current_projects/bioenergy.*

Stanley, B., and R. Bradley. 2008. Plants at the Pump: Reviewing Biofuels' Impacts and Policy Recommendations. World Resources Institute. Climate and Energy Policy Series. Washington, DC.

Yacobucci, Brent. 2006. Fuel Ethanol: Background and Public Policy Issues. Congressional Research Service. Washington, DC.

ENVIRONMENTAL AND HEALTH IMPACTS

Biomass Research and Development Board. 2008. Increasing Feedstock Production for Biofuels: Economic Drivers, Environmental Implications and the Role of Research. *http://www.brdisolutions.com/default.aspx.*

Congressional Budget Office. 2009. The Impact of Ethanol Use on Food Prices and Greenhouse-Gas Emissions. *http://www.cbo.gov/ftpdocs/100xx/doc10057/04-08-Ethanol.pdf.*

Crutzen, P.J., A.R. Mosier, K.A. Smith, and W. Winiwarter. 2008. N2O Release from Agro-Biofuel Production Negates Global Warming Reduction by Replacing Fossil Fuels. *Atmospheric Chemistry and Physics* 8: 389-395.

Dale, Bruce 2008. Biofuels: Thinking Clearly about the Issues. Journal of Agriculture and Food Chemistry 56:3885-3891.

Dominguez-Faus, R., Susan E. Powers, Joel G. Burken, and Pedro J. Alvarez. 2009. The Water Footprint of Biofuels: A Drink or Drive Issue? *Environmental Science and Technology* 43:9: 3005-3010.

Donner, Simon and Christopher Kucharik. 2008. Corn-Based Ethanol Production Compromises Goal of Reducing Nitrogen Export by the Mississippi River. *Proceedings of the National Academy of Sciences* 105:11: 4513-4518.

Fargione, Joseph, Jason Hill, David Tilman, Stephen Polasky, and Peter Hawthorne. 2008. Land Clearing and the Biofuel Carbon Debt. *Science* 319: 1235-1238.

Farrell, Alexander E., Richard J. Plevin, Brian T. Turner, Andrew D. Jones, Michael O'Hare, and Daniel M. Kammen. 2006. Ethanol can contribute to Energy and Environmental Goals. *Science* 311:506-508.

Gerbens-Leenes, Winnie, Arjen Y. Hoekstra, Theo H. van der Meer 2009. The Water Footprint of Bioenergy. PNAS, 106:25:10219-10223.

Gohlke, Julia M., Sharon H. Hrynkow, and Christopher J. Portier. 2008. Health, Economy, and Environment: Sustainable Energy Choices for a Nation. *Environmental Health Perspectives* 116:6: A236-A237.

Groom, Martha J., Elizabeth M. Gray and Patricia A. Townsend 2008. Biofuels and Biodiversity: Principles for Creating Better Policies for Biofuels. Conservation Biology 22:3:602-609.

Hill, Jason, Stephen Polasky, Erik Nelson, David Tilman, Hong Huo, Lindsay Ludwig, James Neumann, Haochi Zheng, and Diego Bonta. 2009. Climate change and health costs of air emissions for biofuels and gasoline. *Proceeding of the National Academy of Sciences* 106:2077-2082.

Hill, Jason, Erik Nelson, David Tilman, Stephen Polasky, and Douglas Tiffany. 2006. Environmental, Economic, and Energetic Costs and Benefits of Biodiesel and Ethanol Biofuels. *Proceedings of the National Academy of Sciences* 103:11206-11210.

Kim, Hyungtae, Seugdo Kim, and Bruce Dale 2009. Biofuels, Land Use Change, and Greenhouse Gas Emissions: Some Unexplored Variables. *Environmental Science and Technology* 43: 3:961-967.

Jacobson, Mark Z. 2007. Effects of Ethanol (E85) Versus Gasoline Vehicles on Cancer and Mortality in the United States. *Environmental Science & Technology* 41:4150-4157.

Landis, Doug, Mary G. Gardiner, Wopke van der Werf, and Scott Swinton. 2008. Increasing Corn for Biofuel Production Reduces Biocontrol Services in Agricultural Landscapes. *Proceedings of the National Academy of Sciences* 20552:105:51.

Mapemba, Lawrence D., Francis M. Epplin, Charles M. Taliaferro, and Raymond L. Huhnke. 2007. Biorefinery Feedstock Production on Conservation Reserve Program Land. *Review of Agricultural Economics* 29:2:227-246.

Marshall, Liz, and Zacery Sugg. 2008. Finding Balance Agricultural Residues, Ethanol, and the Environment. World Resources Institute. Climate and Energy Policy Note, Number 3.

Naidenko, Olga V. 2009. Ethanol-Gasoline Fuel Blends May Cause Human Health Risks and Engine Issues. Environmental Working Group.

National Research Council (NRC). 2008. Water Implications of Biofuel Production in the United States. National Academies Press. Washington, DC.

Renewable Fuels Association. 2008. Understanding Land Use Change and U.S. Ethanol Expansion. Renewable Fuels Association. Washington, DC.

Robertson, G. Philip, Virginia H. Dale, Otto C. Doering, Steven P. Hamburg, Jerry M. Melillo, Michele M. Wander, William J. Parton, Paul R. Adler, Jacob N. Barney, Richard M. Cruse, Clifford S. Duke, Philip M. Fearnside, Ronald F. Follett, Holly K. Gibbs, Jose Goldemberg, David J. Mladenoff, Dennis Ojima, Michael W. Palmer, Andrew Sharpley, Linda Wallace, Kathleen C. Weathers, John A. Wiens, and Wallace W. Wilhelm. 2008. Sustainable Biofuels Redux. *Science* 322:49-50.

Scientific Committee on Problems of the Environment. 2009. Biofuels: Environmental Consequences and Interactions with Changing Land Use.

Searchinger, T., R. Heimlich, R.A. Houghton, Fengxia Dong, A Elbeid, J. Fabiosa, S. Tokgoz, D. Hayes, and Tun-Hsiang Yu. 2008. Use of U.S. Croplands for Biofuels Increases Greenhouse Gases through Emissions from Land Use Change. *Science* 319:1238-1240.

Secchi, Silvia and Bruce Babcock. 2007. Impact of High Crop Prices on Environmental Quality: A Case of Iowa and the Conservation Reserve Program. Working Paper 07-WP 447. Center for Agricultural and Rural Development, Iowa State University. Ames, IA.

Williams, Pamela R.D., Daniel Inman, Andy Arden and Garvin A. Heath 2009. Environmental and Sustainability Factors Associated with Next Generation Biofuels in the U.S.: What Do We Really Know? Environmental Science and Technology 43:13:4763-4775.

Wu. M., M. Mintz, M. Wang, and S. Arora 2009. Consumptive Water Use in the Production of Ethanol and Petroleum Gasoline. Center for Transportation Research, Energy Systems Division, Argonne National Laboratory.

U.K. Renewable Fuels Agency. 2008. The Gallagher Review of the Indirect Effects of Biofuel Production.

SOCIAL/COMMUNITY IMPACTS

Dehue B, S. Meyer, and C. Hamelinck. 2007. Towards a Harmonized Sustainable Biomass Certification Scheme. Ecofys. Commissioned by WWF International. *http://www.* senternovem.*nl/mmfiles/Ecofys%202007%20-%20Harmonised%20Sustainable%20Biomass%20Scheme_tcm24-280157.pdf.*

Dong, Fengxia. 2007. Food Security and Biofuels Development: The Case of China. Briefing Paper: 07-BP 52. October 2007. Center for Agricultural and Rural Development. Iowa State University. Ames, IA.

Iowa State University. 2007. The Bioeconomy in Iowa: Local Conversations. An Iowa State University Extension Report. *http://www.extension.iastate.edu/Publications/SP307.pdf.*

Iowa State University. 2008. 2007 Survey Report on Iowa Farmers' Views on the Bioeconomy. Iowa Farm and Rural Life Poll. *http://www.soc.iastate.edu/extension/presentations/publications/farm/PM2050.pdf.*

Jordan, N., G. Boody, W. Broussard, J. D. Glover, D. Keeney, B. H. McCown, G. McIsaac, M. Muller, H. Murray, J. Neal, C. Pansing, R. E. Turner, K. Warner, and D. Wyse. 2009. Sustainable Development of the Agricultural Bio-Economy. *Science* 316: 1570-1571.

U.S. Department of Agriculture-CSREES. 2007. The Human and Social Dimensions of a Bioeconomy: Implications for Rural People and Places.

Appendix G

Biographical Information: Workshop Participants

PAUL ARGYROPOULOS is the senior policy advisor in the U.S. Environmental Protection Agency (EPA)'s Office of Transportation and Air Quality. Mr. Argyropoulos joined EPA's Office of Transportation and Air Quality's Immediate Office as a senior policy advisor in February of 2006. He is responsible for providing advice and analysis to the office director on a broad range of transportation program issues, with a focus on fuels. For the past 6 years, he worked for Hart Downstream Energy Services serving as executive director of the International Fuel Quality Center, director of Federal Affairs, and executive director of World Refining & Fuels Conferences. Prior to joining Hart, he spent 2 years as a fuels product associate with the American Petroleum Institute, where he provided regulatory and policy coordination among API Downstream Committees on national and state fuel regulatory and policy issues. From 1980 to 1997, Mr. Argyropoulos worked for the EPA in several areas of the agency. However, the majority of his time—more than 13 years—was spent in the Office of Mobile Sources supporting development, implementation, and enforcement of federal and state regulations, including both fuel quality and vehicle emissions controls.

PATRICK ATKINS (Steering Committee Member) recently retired from the position of the director of Technology-Energy Innovation and was responsible for Alcoa's environmental policy and global environmental programs. Dr. Atkins joined Alcoa in Pittsburgh in 1972, after serving as a professor in Environmental Health Engineering at the University of Texas at Austin where he taught engineering, industrial hygiene and ecology courses and directed M.S. and Ph.D. research projects. He became manager-environmental control at Alcoa in 1973, director-environmental control in 1980, director of environmental affairs in 1991 and to his director's position in 1995. He also served as Alcoa's chief environmental engineer from 1982 to 1984. Author of over 50 technical articles and editor of two books, Dr. Atkins is a member of the American Society of Civil Engineers, the National Society of Professional Engineers and the Engineering Society of Western Pennsylvania. He represents Alcoa on the environmental committees of the International Primary Aluminum Institute, the Business Roundtable, National Association of Manufacturers and other national and international groups. In addition, he is a former member of the National Research Council's Commission on Geosciences, Environment and Resources. Dr. Atkins is a registered professional engineer in the states of Texas and Pennsylvania and is an adjunct professor at the University of Pittsburgh Graduate School of Public Health, teaching industrial waste treatment technology. Dr. Atkins received a bachelor's degree in civil engineering from the university of Kentucky in 1964 and master of science in environmental engineering from Stanford University a year later. He also earned a doctor of philosophy degree in 1968 from Stanford specializing in environmental engineering.

BRUCE BABCOCK is a professor of Economics and the Director of the Center for Agricultural and Rural Development at Iowa State University. Professor Babcock's research interests include understanding agricultural commodity markets, the impacts of biofuels on U.S. and world agriculture, the development of innovative risk management strategies for farmers, and the analysis of agricultural and trade policies. Professor Babcock is originally from Southern California. He received his B.S. in economics of resource use and his M.S. in agricultural economics from the University of California at Davis, and his Ph.D. in agricultural and resource economics from the University of California at Berkeley.

CARMEN BAIN is an assistant professor in the Department of Sociology at the Iowa State University (ISU). Dr. Bain's research interests include the political economy of global agri-food systems, international development, and social studies of science and technology. She has conducted research in Chile, Ghana, New Zealand, and the United States. Her work has been published in the journals Rural Sociology and Food Policy and several edited volumes including Agricultural Governance: Globalization and the New Politics of Regulation; Supermarkets and Agri-food Supply Chains and; Between the Local and the Global. Confronting Complexity in the Contemporary Food Sector. Her current research focuses on the social and economic impact of bioeconomy and biofuels development in Iowa. She is the advisor for the Public Service and Administration in Agriculture Program at ISU. Dr. Bain received a bachelor's and master's degree from the University of Canterbury in Christchurch, New Zealand. She also earned a doctor of Philosophy degree from Michigan State University.

PAUL BATCHELLER is a partner at PrairieGold Venture Partners where he oversees all aspects of the firm's investment activities, from sourcing, structuring, and negotiating investments to serving as a board member for portfolio companies. His investment activities at PrairieGold are focused on Greentech and IT. He currently serves on the boards of iCentera, Game Plan Technologies, Augusta Systems, and a cellulosic ethanol company that is currently in stealth mode. In addition, he oversees PrairieGold's investment in General Compression, a developer of next-generation wind turbines. Mr. Batcheller is also a board member of South Dakota Rural Enterprise, a statewide non-profit economic development organization. Prior to joining PrairieGold, he served for 7 years as an advisor to Senator Tom Daschle, where he worked on economic policy, science, and technology issues. He received his B.A. in economics from Macalester College.

MICHAEL BELL is an associate professor of Rural Sociology at the University of Wisconsin-Madison. He is principally an environmental sociologist, but he also conducts research on culture, economic sociology, sustainable agriculture, community, place, rural society, inequality, gender, the body, democracy, and social theory. Two central themes can be heard in all of his work: dialogics and the sociology of "nature," broadly conceived. He is the author of Childerley: Nature and Morality in a Country Village (University of Chicago Press, 1994), which was co-winner of the 1995 Outstanding Book Award of the Sociology of Culture Section of the American Sociological Association. Along with Gregory Peter, Susan Jarnagin, and Donna Bauer, he is the author of the forthcoming book Farming for Us All: Practical Agriculture and the Cultivation of Sustainability (Pennsylvania State University Press, 2004). The second edition of his An Invitation to Environmental Sociology (Pine Forge Press [Sage]), 1998) appeared in 2004. Dr. Bell has also worked as a geologist, and is the author of The Face of Connecticut: People, Geology, and the Land (State of Connecticut, 1985), which won an American Library Association award. He continues to have a second life as a part-time composer of songs, fiddle tunes, and classical music. He also plays mandolin in an old-time string band, the Barn Owl Band, which recently appeared on the national public radio show A Prairie Home Companion. He is currently at work on a string quartet, a suite for piano, and a symphonic poem.

MARILYN BUFORD provides national leadership for U.S. Forest Service research programs in silviculture and sustainable forest productivity, and co-leads the FS Biobased Products and Bioenergy Research Program. Dr. Buford joined the Forest Service R&D National Program Staff in August, 1998, serving as national program leader for Quantitative Ecology Research and co-lead for Forest Service Global Change Research Program. She served as scientist and project leader in Charleston, SC (Forested Wetlands) and in Research Triangle Park, NC (Southern Forest Productivity) from 1985 to 1998. During that time, she helped form and lead the Short Rotation Woody Crops Cooperative Research Program located at the Savannah River Site (Aiken, SC). She is a leader of the U. S. Department of Agriculture Biobased Products and Bioenergy Coordination Council, an active member of the Interagency Woody Biomass Utilization Group, and serves on numerous interagency teams providing analysis and technical information for federal bioenergy and biobased products efforts. Dr. Buford is immediate past chair of the Short Rotation Woody Crops (SRWC) Operations Working Group, a public-private partnership to promote collaborative efforts in developing needed science and technology for SRWC plantations. Her personal research and publications have focused on forest stand dynamics, forest carbon management, and forest productivity. She earned a B.S. in biology from Rhodes College (Memphis, TN), M.S. in silviculture (State University of New York College of Environmental Science and Forestry) and Ph.D. in forestry from North Carolina State University. She is a native of Houston, TX.

JOHN CARBERRY (Steering Committee Member) recently retired from the position of the director of Environmental Technology for the DuPont Company in Wilmington, DE. He was responsible for recommendations on technical programs for DuPont based on an analysis of environmental issues. He led this technology function in a transition to increasingly emphasize waste prevention and product stewardship while maintaining excellence in treatment. Externally, Mr. Carberry is a past chair of the standing National Academy Committee on the Destruction of the Non-Stockpile Chemical Weapons, a founding member of the Green Power Market Development Group and of the Vision2020 Steering Committee, and a member of the National Academy of Engineering Committees on; Technologies for Sequestering CO_2, and Metrics for Documenting Progress in Global Change Research. Since 1990, he has served on four other National Academies' Committees and has presented 30 lectures on environmental issues at 18 universities, given invited presentations at 63 public conferences worldwide and provided 21 literature interviews, or contributions. He holds a B.ChE. and an M.E. in Chemical Engineering from Cornell University and an MBA from the University of Delaware.

PETER CIBOROWSKI (Steering Committee Member) leads the climate change team of the Minnesota Pollution Control Agency. He has served on the steering committees and working groups of the University of Minnesota's Terrestrial Carbon Sequestration Initiative, Sustainable Transportation Initiative, and adaptation initiative. He represented the Midwest in the discussions leading to the design of the Climate Registry (TCR) and has served on the TCR General Reporting Protocol working group, the TCR Mandatory Reporting Committee, and working groups developing reporting protocols for the electric power sector and local governments. The TCR is a 42-state effort to develop a uniform national reporting system for GHGs. He served on the Midwest Registry committee and the USEPA Emission Inventory Improvement Program for GHGs and was an invited presenter to the White House Task Force on Climate Change under the Clinton Administration. He directs the work of MPCA staff on the model rule and reporting and standards committees of the Midwest Accord, the six-state

Midwest effort to develop a cap-and-trade program for GHGs. He is author of the Minnesota Climate Action Plan and, just recently, the 2009 MPCA report on "Minnesota GHG Emissions, 1970-2006: Update and Progress Report." He developed the analysis for Minnesota's GHG externality value for energy planning and Minnesota's environmental disclosure reporting, as well as the analysis of statewide GHG emissions used for the Minnesota Climate Change Advisory Group (MCCAG), the Governor's 2007-2008 GHG task force. He sat on the MCCAG emission inventory, energy supply and cross-cutting committees. Recent projects include: the MPCA guidance incorporation of GHGs into state environmental review and permitting processes and the MPCA solid waste stakeholder process for reducing GHG emissions. He holds a masters degree in Public Affairs from the University of Minnesota.

RANDALL FORTENBERY is the RENK Chair in Agribusiness, Agricultural and Applied Economics and the director of Renk Agribusiness Institute in the Department of Agricultural and Applied Economics at the University of Wisconsin-Madison. His research program currently focuses on agricultural price performance in local and national markets. He is also engaged in studying the impact of futures price action on the stability of cash prices. Another part of his research is identifying specific causal effects of recent price action in agricultural markets. This research includes the impact of U.S. futures trading on the price structure in the developing countries of Central America. Dr. Fortenbery holds a B.S. in Natural Resource Economics and an M.S. in Applied Economics from Montana State University, and a Ph.D in Agricultural Economics from the University of Illinois-Urbana/Champaign.

ALISA GALLANT is a research physical scientist and ecosystems geographer with the U.S. Geological Survey's Earth Resources Observation and Science (EROS) Center. She conducts multiscale, integrated, geospatial research to characterize the landscape and landscape change and to assess the consequences of change on ecosystem functions and sustainability with respect to wildlife and biodiversity. She is a principal investigator on an interdisciplinary team estimating the effects of alternative landscape futures, as driven the demand for biomass for energy and a shifting climate, on ecosystem processes and services in the northern Great Plains. Dr. Gallant holds B.A.'s in Biology and Art from Sonoma State University, an M.S. in Biological Science from Oregon State University, and a Ph.D. through a Remote Sensing and GIS program from Colorado State University.

ALISON GOSS ENG is currently the lead for sustainability research and development programming for the Biomass and Bioenergy Program at the U.S. Department of Energy. She received her Ph.D. from Purdue University in Earth and Atmospheric Sciences, and has a background in terrestrial ecology, hydrogeography and human impacts on water resources. She also currently serves on the Interagency Sustainability Working Group under the Federal Biomass Research and Development Board. Dr. Eng is a member of the U.S. delegation on the Global Bioenergy Partnership's Greenhouse Gas Accounting and Sustainability working groups, and represents the Department of Energy on the Council for Sustainable Biomass Production, a multi-stakeholder group developing biomass to biofuel sustainability principles and standards for the production of feedstocks for second-generation biorefineries. She is also representing the United States as a lead author on the bioenergy chapter of a new International Panel on Climate Change (IPCC) report on renewable energy and climate change mitigation.

ELISABETH A. GRAFFY (Steering Committee Member) has worked on national, state, and international sustainability policies and programs for more than 20 years, and as policy advisor and economist with the U.S. Geological Survey for the last decade. She most recently served as the Department of the Interior's Coordinator for Environmental Indicators and representative on the interdepartmental team that designed the national environmental indicators initiative, announced by the White House in 2008. She participates in the federal interagency effort to develop sustainability indicators for biofuels and is collaborating with partners in state and federal agencies, universities, and non-governmental organizations to explore frameworks for addressing energy and other cross-cutting issues, with particular attention to the special challenges of bridging research and policy development. During her tenure with the U.S. Congress Office of Technology Assessment, Dr. Graffy co-authored two major policy assessments on agriculture, environment, and trade. While at USGS, she has authored, edited, or overseen numerous publications on related themes and developed new conceptual models related to the intersection of science and policy. Her papers and public presentations appear in many forums, including conference proceedings and journals such as *Society and Natural Resources*, the *International Journal of Global Environmental Issues*, and *Public Administration Review*. She holds an A.B. in Politics from Princeton University, an M.S. in Agricultural Economics from the University of Wisconsin-Madison, and a Ph.D. in Policy from the Gaylord Nelson Institute for Environmental Studies at the University of Wisconsin-Madison.

NATHANAEL GREENE (Steering Committee Member) is a senior policy analyst of the Natural Resources Defense Council. Greene received his Bachelor of Arts Degree in Public Policy from Brown University and a Master of Science Degree in Energy and Resources from University of California Berkeley. He joined NRDC in 1992 and worked 2 years before getting his master's degree and returned to NRDC in 1996 and working there since. He is a senior policy analyst and is responsible for working on energy policy and related issues including utility restructuring, energy taxes, energy efficiency, renewables, and low-income services. He has particular expertise in clean energy technologies including wind, solar and biomass energy, fuel cells, combined heat and power and energy efficiency and in regulations and policies to promote these technologies. For the past few years he has been focusing on assessing the sustainable potential for biofuels and developing policies to advance them.

ALAN HECHT is the director for Sustainable Development in the U.S. Environmental Protection Agency (EPA)'s Office of Research and Development. He was Associate Director for Sustainable Development at the White House Council on Environmental Quality (2002-2003) and Director of International Environmental Affairs for the National Security Council (2001-2002). He served as the White House coordinator for the 2002 World Summit on Sustainable Development. He was the Deputy Assistant Administrator for International Activities at the EPA (1989-2001). Twice he received EPA's highest award, the Gold Medal, for leading U.S. negotiations for the environmental side agreement to the North American Free Trade Agreement and for his innovative work on promoting nuclear waste management in Russia. He has recently published articles on sustainable development in Environmental Forum (2003) and Water Policy (2004). Dr. Hecht earned a Ph.D. degree at Case Western Reserve University.

JASON HILL (Steering Committee Member) is a research associate in the Department of Applied Economics and the Department of Ecology, Evolution, and Behavior at the University of Minnesota. His research interests include the technological, environmental, economic, and social aspects of sustainable bioenergy production from current and next-generation feedstocks. His work on the life cycle impacts of transportation biofuels has been published in the journals Science and the Proceedings of the National Academy of Sciences. He is currently focusing on the effects that the expanding global biofuels industry is having on climate change, land use, biodiversity, and human health. Dr. Hill has testified before U.S. Senate committees on the use of diverse prairie biomass for biofuel production and on the greenhouse gas implications of ethanol and biodiesel. He has also performed independent analysis for the National Renewable Energy Laboratory, the National Research Council, and the U.S. Environmental Protection Agency. Dr. Hill received his A.B. in biology from Harvard College and his Ph.D. in plant biological sciences from the University of Minnesota.

TRACEY HOLLOWAY (Steering Committee Member) is the director of the Center for Sustainability and the Global Environment (SAGE), a cross-disciplinary research center based in the Nelson Institute for Environmental Studies at the University of Wisconsin-Madison. Dr. Holloway's research examines air pollution chemistry and transport at regional and global scales, including links between air quality and climate, energy, land use, health, and public policy. As an assistant professor in the Nelson Institute, she teaches graduate and undergraduate courses on environmental modeling, air resource management, and atmospheric chemistry, and she has affiliate appointments in Atmospheric and Oceanic Sciences (AOS), Civil and Environmental Engineering (CEE), and the La Follette School of Public Policy. Dr. Holloway earned her Ph.D. in AOS from Princeton University in 2001, and completed a certificate in Science, Technology, and Environmental Policy from the Woodrow Wilson School of Public and International Affairs. Her undergraduate degree (Sc.B.) is from Brown University in Applied Mathematics, and her post-doctoral work was done at Columbia University's Earth Institute.

MOLLY JAHN serves as dean of the College of Agricultural and Life Sciences at UW Madison and Director of the Wisconsin Agricultural Experiment Station. Her efforts were instrumental in securing the Department of Energy Great Lakes Bioenergy Research Center on the UW Madison campus and in launching the Wisconsin Bioenergy Initiative. She has worked to coordinate university-based research, extension and outreach in bioenergy with state and federal initiatives and priorities and to support coordinated regional efforts in the Midwest. She serves as the lead dean in the hire of eight new faculty positions committed by the State of Wisconsin to support sustainable bioenergy technologies and for the construction of a $50M facility for sustainable and renewable energy. She also holds a faculty appointment in the Departments of Genetics and Agronomy. Dr. Jahn's research has focused on the genetics, genomics and breeding of crop plants, releasing more than two dozen crop varieties currently grown commercially on six continents. She has also worked extensively overseas to link crop breeding objectives to improvement in human nutrition and income, and currently is active in a number of leadership roles in international development. Dr. Jahn received her B.A. with Distinction in Biology from Swarthmore College and holds graduate degrees from Cornell and MIT. She served 15 years on the faculty at Cornell University prior to assuming her current position.

BRENDAN JORDAN is the program manager of the Great Plains Institute. Mr. Jordan focuses on staffing the Midwestern Governors Association (MGA) Energy Security and Climate Stewardship Platform, the North Central Bioeconomy Consortium (NCBC), and the Native Grass Energy Initiative. His work promotes the development of a Midwestern bioeconomy that stimulates rural economic development, makes improvements to air, soil, and water quality, and addresses global warming. He has a Masters Degree in Science, Technology, and Environmental Policy from the University of Minnesota, and a B.A. in biology from Carleton College.

JIM KLEINSCHMIT is the director of Rural Communities Program for the Institute for Agriculture and Trade Policy (IATP). Kleinschmit's work focuses on promoting working landscapes and sustainable rural development in both the United States and abroad. Current projects include: working with farmers and other stakeholders to establish sustainable crop production standards and markets in the Midwest; promoting and facilitating renewable energy and sustainable bioindustrial development projects; and helping increase understanding of the relationship of agriculture to surface and ground water management in the Great Lakes Basin. He has a M.A. from the Jackson School of International Studies of the University of Washington, and a B.A. in European history and Russian studies from St. Olaf College, Minnesota. He was raised on and is still active in the operation of his family's farm in Nebraska. He worked on rural development in the Baltics and Russia and in 1995 began working as the coordinator for the IATP's International Fellows Program, which informed officials from the former Soviet Union and Eastern Europe about international trade and agriculture issues. In 1996, he joined the Environment and Agriculture Program, focusing on nutrient and watershed management.

PATRICIA KOSHEL (Staff) is a senior program officer with the National Academies' Policy and Global Affairs Division. She has been the staff lead for a consensus study on science and technology in U.S. Foreign Assistance Programs and has also worked on the Science and Technology for Sustainability Program. Before joining the National Academies, Ms. Koshel was the director of Bilateral Programs in the Office of International Affairs at the U.S. Environmental Protection Agency. Before that she was the Energy and Environmental Policy Advisor for the U.S. Agency for International Development. She has a master's degree in economics.

CHRIS KUCHARIK is an assistant professor of Agronomy and Environmental Studies at the University of Wisconsin-Madison. He graduated from the University of Wisconsin-Madison in 1997 with a Ph.D. in Atmospheric Sciences (minor soil science). During his graduate studies, he participated in the BOReal Ecosystem-Atmosphere Study (BOREAS), an international field experiment that took place in the Canadian boreal forest. He helped design a high-resolution, two-band, ground-based remote-sensing instrument, called a Multiband Vegetation Imager—which allowed for advanced studies of forest canopy architecture and enabled for more accurate predictions of carbon cycling in high latitude ecosystems. Currently, his research focuses on integrating field observations and numerical models of natural and managed ecosystems to better understand the influence of changing climate and land management on ecosystem services. Dr. Kucharik's interests include carbon cycling and sequestration in wetlands, prairie ecosystems, and agricultural landscapes, water quality, biofuels, and how crop yields are affected by climate change and farmer management. This work has been supported by a NASA Interdisciplinary Science (IDS) grant, through the DOE National Institute for Climate Change Research (NICCR), Madison Gas and Electric, S.C. Johnson, and a Wisconsin Focus on Energy grant.

KATHLEEN McALLISTER (Staff) is a research associate with the Science and Technology for Sustainability Program (STS) at the National Academies. Before joining the National Academies in 2006, she attended Lehigh University and graduated with highest honors with a B.A. in Sociology. Ms. McAllister wrote an honors thesis on social implications of human trafficking in the United States and worked throughout her college career as a research assistant for professors of Sociology at Lehigh University. She also speaks conversational Spanish, and has had internships in the offices of U.S. Representative Paul E. Kanjorski and U.S. Senator Arlen Specter. She is concurrently pursuing her M.S. in Environmental Science and Policy at Johns Hopkins University.

JOHN A. MIRANOWSKI (Steering Committee Member) is a professor in the Department of Economics at Iowa State University. He served as chair of the department from 1995 to 2000. Dr. Miranowski has expertise in soil conservation, natural-resource management, water quality, land management, energy, global change, and agricultural research decision making. He has previously served as director of the Resources and Technology Division of the U.S. Department of Agriculture Economic Research Service (1984-1994); executive coordinator of the secretary of agriculture's Policy Coordination Council and special assistant to the deputy secretary of agriculture (1990-1991); and Gilbert F. White fellow at Resources for the Future (1981-1982). Dr. Miranowski headed the U.S. delegation to the Organization for Economic Cooperation and Development Joint Working Party on Agriculture and the Environment (1993-1995). He has served as a member of the Ad Hoc Working Group on Risk Assessment of Federal Coordinating Committee on Science, Education, and Technology (1990-1992); director of the Executive Board of the Association of Environmental and Resource Economists (1989-1992); and director of the Executive Board of the American Agricultural Economics Association (1987-1990). Dr. Miranowski is currently serving on the Alternative Liquid Transportation Fules Committee of the National Research Council's America's Energy Future Study. He served as a member of the National Research Council Committee on Impact of Emerging Agricultural Trends on Fish and Wildlife Habitat. He received a B.S. degree in agricultural business from Iowa State University and M.A. and Ph.D. degrees in economics from Harvard University.

MARINA S. MOSES (Staff) recently joined the Policy and Global Affairs Division of the National Academies as the Director for the Science and Technology for Sustainability Program. Prior to joining the Academies, Dr. Moses served on the faculty of The George Washington University School of Public Health and Health Services in the Department of Environmental and Occupational Health. At the University, Dr. Moses was the director of the Doctoral Program and the Practicum Coordinator for the graduate program. Dr. Moses was the recipient of the 2005 Pfizer Scholar in Public Health Award and has worked in emergency preparedness and communication with communities on public health issues. Currently, she is the president of National Capital Area Chapter of the Society of

Risk Analysis. Before joining the faculty at the George Washington University, Dr. Moses held senior scientific positions in the Environmental Management Division of the U.S. Department of Energy (DOE) and in the Superfund Program of the U.S. Environmental Protection Agency (EPA) in a Regional office. At the DOE, she worked on the development of a qualitative framework to assess hazardous and nuclear waste risks from DOE sites and helped establish a long-term research program on "transformational" science. Prior to her experience at DOE, she served as the senior human health risk assessor in the New York City Office of EPA's Superfund Program where she worked on risk assessments that addressed abandoned hazardous waste sites and developed national guidance and policies in this area. During her years in New York City, she also held an appointment as Assistant Adjunct Clinical Professor of Public Health in the Columbia University College of Physicians and Surgeons. Dr. Moses received her B.A. (Chemistry) and her M.S. (Environmental Health Sciences) degrees from Case Western Reserve University. She received her Doctorate of Public Health (Environmental Health Sciences) from Columbia University School of Public Health.

MARCIA PATTON-MALLORY (Steering Committee Member) is a loaned executive from the U.S. Department of Agriculture Forest Service. She works closely with the Western Forestry Leadership Coalition to help accomplish their strategic goals related to biomass utilization, bioenergy, and climate change. She also is a member of the Forest Service's Climate Change Strategy team working on mitigation, and participates with regional and national climate change initiatives in relation to forestry and bioeneryg. Previously, she coordinated the woody biomass efforts of the USDA Forest Service across programs and provided executive liaison and coordination between the USDA Forest Service and other federal agencies, state organizations and private interests. She has 25 years of Forest Service experience as: station director of the Rocky Mountain Research Station, Fort Collins, CO; staff specialist in Forest Products and Harvesting Research, Washington, DC; and research engineer, Forest Products Laboratory, Madison, WI. Additional relevant experience includes Science and Technology Fellow in the U.S. Senate working on energy and natural resources issues, and internships with Weyerhaeuser Company, Tacoma, WA.

GREG NEMET is an assistant professor at the University of Wisconsin in the La Follette School of Public Affairs and the Nelson Institute for Environmental Studies. He is also a member of the university's Energy Sources and Policy Cluster and a senior fellow at the university's Center for World Affairs and the Global Economy. His research and teaching focus on improving understanding of the environmental, social, economic, and technical dynamics of the global energy system. He teaches courses in international environmental policy and energy systems analysis. A central focus of his research involves empirical analysis of the process of innovation and technological change. He is particularly interested in how the outcomes of this line of research can inform public policy related to improvements in low-carbon energy technologies. His work is motivated by a more general interest in issues related to energy and the environment, including how government actions can expand access to energy services while reducing their environmental impacts. He is a lead author of the Global Energy Assessment. He holds a master's degree and doctorate in energy and resources, both from the University of California, Berkeley. His undergraduate degree from Dartmouth College is in geography and economics.

PETER NOWAK served as both an assistant and associate professor at Iowa State University before joining the faculty at the University of Wisconsin in 1985. At the College of Agricultural and Life Sciences in Madison he holds multiple appointments as a Soil and Water Conservation Specialist in the Environmental Resources Center, Research Professor in the Department of Rural Sociology, and Chair of Academic Programs in the Gaylord Nelson Institute for Environmental Studies. He also served as Chair of the Wisconsin Buffer Initiative for the last three years. Pete's career has focused on measuring and explaining the adoption and diffusion of agricultural technologies, especially those with natural resource management implications. More recently he has focused on examining the application of spatial analytical techniques and statistics to critical issues in resource management. His work has been published in a variety of journals and books. He has served as an Associate Editor for the Journal of Soil and Water Conservation, Editorial Board of the Journal of Precision Agriculture and on the Foundation for Environmental Agricultural Education. In the recent past he has worked with the National Academy of Science's Board on Agriculture, U.S. Army Corps of Engineers, U.S. Office of Management and Budget, USDA's Natural Resources Conservation Service and a National Blue Ribbon Panel examining the USDA Conservation Effectiveness Assessment Project. He also served on the Board of Directors of the Soil and Water Conservation Society. He received his Ph.D. from the University of Minnesota's College of Agriculture in 1977.

DONNA PERLA is a senior advisor in the Office of Research and Development at the U.S. Environmental Protection Agency. She leads the Office of Research and Development's biofuels effort and assists EPA's representative to the federal Biomass Research and Development Board and participates in several interagency teams related to the development of a National Biofuels Action Plan. Her work focuses on looking at the sustainability of the biofuels system, including environmental and human health considerations of feedstock, technologies, distribution and use. Donna also leads an EPA Waste-to-Energy network, which explores the environmental aspects of conversion technologies for a wide variety of wastes, including disaster debris. Other positions in her 22 years with EPA include: director of the Innovative Pilots Division in the Office of Policy, Economic, and Innovation; chief of the Waste Minimization Branch in the Office of Solid Waste, Chief of the Colorado/Montana Permitting and Enforcement Section, EPA, Region 8; chief of the Economic Analysis and Risk Assessment Section in the Office of Solid Waste; and special assistant to the Director of the Office of Solid Waste. She holds a B.S. in Biology (University of Hartford) and a Masters of Public Health (Yale University).

GARY RADLOFF (Steering Committee Member) is the director of Policy and Strategic Communications at the Wisconsin Department of Agriculture, Trade and Consumer Protection shaping department-wide policy initiatives and communication plans. He is staff liaison to the North Central Bioeconomy Consortium (NCBC), a 12-state partnership of Agriculture departments, University Extension offices and Agriculture Research Stations. Radloff serves on the Steering Committee for the Midwest Agriculture Energy Network (MAEN), a regional policy incubator. He is on the Advisory Committee to the Wisconsin Initiative on Climate Change Impacts (WICCI), researching and providing outreach on climate change adaptation policy and practices. Recent projects in promoting renewable energy policy and climate stewardship include advising the Agriculture and Forestry Work Group of the Governor's Task Force on Global Warming. He also assisted with policy planning and platform development for the Midwest Governor's Association, Energy Security and Climate Stewardship held in November 2007. Previously, he served as a policy staff and co-author of final reports for two major Wisconsin projects; Governor (Jim Doyle's) Consortium on the Biobased Industry and the Working Lands Initiative. The Governor's Consortium is a roadmap for positioning Wisconsin to play a key role in promoting the use of renewable energy and growing the state's bioeconomy. The Working Lands Initiative is a report of detailed policy steps and strategies to protect the source of food and fiber, paper and pulp, and biomass—the Wisconsin working lands in agriculture and forestry. He is a grant reviewer with the Environmental and Economic Research and Development Program with the Focus on Energy Program, Public Service Commission, and a member of the Universal Service Council of the Public Service Commission. He has a Master's Degree in Public Administration and Public Policy.

JOHN REGALBUTO is currently the director of the Catalysis and Biocatalysis Program in the Engineering Directorate at the National Science Foundation. He is the lead co-chair of the Biomass Conversion Interagency Working Group, which reports to the National Biomass R&D Board. Dr. Regalbuto's home institution is the University of Illinois at Chicago, where he is a professor in the Department of Chemical Engineering. His education includes a B.S. in Chemical Engineering from Texas A&M University in 1981, an M.S. in Chemical Engineering from the University of Notre Dame in 1983 and a Ph.D. from Notre Dame in 1986. Directly thereafter he joined the University of Illinois at Chicago. He has several hundred research publications and presentations, and most recently has edited one of the few books in his research specialty, catalyst preparation. Dr. Regalbuto has twice served as president of the Catalysis Club of Chicago, and has been active organizing symposia on catalysis for meetings for the American Institute of Chemical Engineers and the American Chemical Society. He has 3 children and his wife also holds a Ph.D. in Chemical Engineering.

PHIL ROBERTSON is Professor of Ecosystem Science in the Department of Crop and Soil Sciences at Michigan State University (MSU), with which he has been associated since 1981. Since 1988 he has directed the NSF Long-Term Ecological Research (LTER) Program in Agricultural Ecology at the W.K. Kellogg Biological Station, where he is a resident faculty. He currently serves as chair of the U.S. LTER Network's Science Council and Executive Board. He is also program leader for sustainability in the Department of Energy's Great Lakes Bioenergy Research Center. Dr. Robertson's research interests include the biogeochemistry and ecology of field crop ecosystems, including biofuel systems, and in particular nitrogen and carbon dynamics, greenhouse gas fluxes, and the functional significance of microbial diversity in these systems. Dr. Robertson has been a SCOPE-Mellon postdoctoral fellow at the Royal Swedish Academy of Sciences (1980-1981) and a sabbatical scholar at Cooperative Research Centres in Adelaide (1993-1994) and Brisbane (2001-2002), Australia. His service also includes past membership on the U.S. Carbon Cycle Scientific Steering Committee, chairmanships of competitive grants panels at the USDA (the NRI and Fund for Rural America Programs), and membership on several NSF panels in the Biological and Geosciences directorates. He served on the National Research Council Committee to Evaluate the USDA NRI Program (1998-1999), and chaired the Environment Subcommittee of the NRC Committee on Opportunities in Agriculture (2000-2002). He has testified before the U.S. Senate Agriculture, Forestry, and Nutrition Committee and participated in briefings for the U.S. House Science and Agriculture Committees. He has also served as an editor for the journals Ecology, Ecological Monographs, and Plant and Soil and is currently an editor for Biogeochemistry. In 2003, he was elected a Fellow in the Soil Science Society of America. In 2005 he received MSU's Distinguished Faculty award. Dr. Robertson received his B.A. from Hampshire College and his Ph.D. in Biology from Indiana University.

BRUCE D. RODAN (Steering Committee Member) is a Senior Policy Advisor-Environment in the White House Office of Science and Technology Policy (OSTP). Dr. Rodan serves as OSTP liaison to the Ecosystems and the Toxics and Risk Subcommittees of the NSTC Committee on Environment and Natural Resources (CENR). Dr. Rodan is a medical doctor (U. Melb) with Masters Degrees in Environmental Studies (U. Melb) and Public Health (Harvard). His work has included environmental risk analyses for toxic chemicals under the U.S. EPA

Integrated Risk Information System (IRIS), negotiating the Stockholm Convention on Persistent Organic Pollutants (POPs), and research on neotropical timber species under the CITES Treaty.

RUTH SCOTTI is the Biofuels Regulatory Affairs Manager for BP Biofuels. She constructs advocacy strategy and company advocacy positions for BP's new Biofuels business. While at the University of Michigan, she was a summer associate in the renewable energy leadership program at GE Wind. Prior to that she conducted market research in Taiwan and funding strategies for U.S. grant makers seeking to fund Chinese non-governmental organizations. She holds an undergraduate degree in psychology and biology with minors in chemistry and Asian studies. She is fluent in Mandarin Chinese and speaks conversational French.

THERESA SELFA, assistant professor of Sociology, has expertise in rural, environmental, agricultural and development sociology, with research experience in Brazil, Philippines, Europe and the US. She was a post-doctoral associate in Washington State on a project examining alternative agriculture and food systems. She recently completed research examining environmental attitudes and behaviors toward land management in Devon, England. Dr. Selfa is currently working as the lead social scientist on an interdisciplinary water quality project assessing impacts of farmers' management behavior on water quality in an agricultural watershed in Central Kansas, and as the lead social scientist in a new interdisciplinary program in Agricultural Resource and Environment Management. She is the PI on a DOE-funded study on the Impacts of Biofuels on Rural Communities in Kansas and Iowa. Her work has been published in *Society and Natural Resources, Environment and Planning A, Journal of Rural Studies*, and *Agriculture and Human Values*. She has a Ph.D. in Development Sociology from Cornell University.

JOHN SHEEHAN serves as the scientific program coordinator for biofuels and the global environment at the University of Minnesota's Institute on the Environment, focusing in particular on direct and indirect consequences of biofuel production on land use across the world. Sheehan has 25 years of experience in chemical engineering, analysis and planning, including 14 years working with biomass technologies. Most recently, he served as vice president of strategy and sustainable development at LiveFuels Inc., a venture capital-funded startup based in California that focuses on algal fuels technology. Prior to that, Sheehan spent nearly two decades with the National Renewable Energy Laboratory, where he conducted pioneering work on system dynamic models for strategic and policy decision-making related to biofuels. During that time, he led the Department of Energy's assessment of its energy efficiency and renewable energy technology portfolio; conducted landmark studies of energy, air quality, greenhouse gas and soil impacts of stover-to-ethanol; oversaw multidisciplinary teams of scientists and engineers; and published numerous peer-reviewed articles on the gamut of energy and environmental topics.

EMMY SIMMONS serves as co-chair of the National Academies' Roundtable on Science and Technology for Sustainability. She is currently an independent consultant on international development issues, with a focus on food, agriculture, and Africa. She serves on the boards of several organizations engaged in international agriculture and global development more broadly: the Partnership to Cut Hunger and Poverty in Africa, the International Livestock Research Institute (ILRI), the International Institute for Tropical Agriculture (IITA), the Washington chapter of the Society for International Development (SID), and the Africa Center for Health and Human Security at George Washington University. Ms. Simmons co-chairs the Roundtable on Science and Technology for Sustainability at the National Academies of Science and leads a Roundtable working group on Partnerships for Sustainability. She completed a career of nearly 30 years with the U.S. Agency for International Development (USAID) in 2005, having served since 2002 as the Assistant Administrator for Economic Growth, Agriculture, and Trade, a Presidentially-appointed, Senate-confirmed position. Prior to joining USAID, she worked in the Ministry of Planning and Economic Affairs in Monrovia, Liberia and taught and conducted research at Ahmadu Bello University in Zaria, Nigeria. She began her international career as a Peace Corps volunteer in the Philippines from 1962-64. She holds an M.S. degree in agricultural economics from Cornell University and a B.A. degree from the University of Wisconsin-Milwaukee.

JEFFERY STEINER is national program leader for Agricultural System Competitiveness and Sustainability with the USDA, Agricultural Research Service–Office of National Programs in Beltsville, MD. He leads nineteen research projects around the country that are producing new kinds of technology and systems to help producers respond to changing environmental and market conditions, enhance natural resources quality, and increase American food, fiber, and energy security. Jeff is also a member of the USDA Council for Sustainable Development, and represents ARS and the USDA Research, Education, and Economics mission area in other matters related to sustainability, particularly in the emerging area of agricultural based bioenergy production. He also coordinates the ARS organic agriculture portfolio. Jeff received his B.S. and M.S. degrees from California State University-Fresno, and the Ph.D. from Oregon State University. He is a fellow of the American Society of Agronomy and Crop Science Society of America.

DAVID SWENSON is an associate scientist in Economics and a lecturer in Community and Regional Planning at Iowa State University, and a Lecturer in the

Graduate Program in Urban and Regional Planning at the University of Iowa. He has an M.A. in urban and regional planning from University of Iowa and an M.A. in political science from University of South Dakota. He teaches planning methods and techniques, urban economics, project evaluation methods, and economic impact assessment. His primary area of research focuses on regional economic changes and their fiscal and demographic implications for communities and local governments in Iowa and in the Midwest. He has developed protocols and conducts targeted industry research for assisting in regional economic development. Mr. Swenson has completed numerous economic impact studies and written and presented extensively about the appropriate methods and interpretations for applying impact analyses to public policies.

DOUGLAS TIFFANY is an extension educator, Agricultural Business Management in the Department of Applied Economics at the University of Minnesota. Current research projects include analysis of production economics of ethanol and biodiesel. Patterns of energy usage by agricultural enterprises as well as emissions of greenhouse gases and the potential for carbon sequestration are continuing interests as well as international climate change treaties. For the year 2001-2002 he was awarded the Endowed Chair in Agricultural Systems by the College of Agricultural, Food, and Environmental Sciences at the University of Minnesota. Much of his research work over the past 10 years has involved analysis of energy production from agriculture as well as the levels of energy required to produce various agricultural products. Working with others, he has analyzed the impact of the Kyoto Accord on Midwestern agriculture and the cost effectiveness of various phosphorous abatement strategies. Livestock consumption patterns and trends of Minnesota crops have been studied as well as the transportation patterns of grains grown in the state. Decision-making tools have been developed by him through the years for ethanol plant operators, farmers considering precision agricultural technology, mining engineers trying to reduce diesel emissions, appraisers needing to discount contract for deed land transfers, and swine farmers seeking to select rations that maximize profits. Mt. Tiffany majored in agricultural economics at the University of Minnesota with a heavy emphasis on the agricultural sciences of agronomy, soils, and animal nutrition. He continued his interest in these areas with more attention to institutional aspects of production while attaining a M.S. degree from the same department. After graduation he worked in state government and in commercial banking for over a decade with most activity in appraisal and valuation of farmland. In addition, he has worked full-time as a self-employed farmer raising agronomic and vegetable crops. He joined the University of Minnesota staff in 1994.

LEANN M. TIGGES is professor of Rural Sociology at the University of Wisconsin in Madison. Her research interests include economic change and labor force issues. She has conducted research on Wisconsin's corn ethanol producers with a special interest identifying the community benefits and costs of hosting an ethanol refinery. Professor Tigges has also conducted research on Wisconsin manufacturers' labor utilization strategies and their global competitive position. She teaches courses on gender, work, and local labor markets. She holds a Ph.D. in Sociology from the University of Missouri.

JOHN YUNKER is a program evaluation coordinator for the Minnesota Office of the Legislative Auditor. His evaluation work has covered government programs in many different areas, including education, transportation, economic development, environmental protection, and health care. In recent years, his work has resulted in major reforms in the operation of the Minnesota State Lottery and in the Jobs Opportunity Building Zone (JOBZ) program, the state's largest economic development program. In April 2009, he authored a report on Minnesota's biofuel policies and programs, which provided an extensive review of the literature on the energy, environmental, and economic impacts of corn-based ethanol. Over the past 30 years, he has testified extensively to legislative committees in Minnesota and worked with executive branch agencies to implement evaluation recommendations. Mr. Yunker received his B.A. in economics from Lawrence University (Wisconsin.) and his M.A. in economics from the University of Minnesota.

JUDY ZIEWACZ is the director of the Wisconsin Office of Energy Independence (OEI) which was created by Governor Doyle on April 5, 2007. Ms. Ziewacz has 32 years of experience in the public and private sectors. Prior to OEI, she served as Deputy Secretary of the Wisconsin Department of Agriculture, Trade and Consumer Protection (DATCP) for 4r years. She served as chief of staff to a Wisconsin Congressman in Washington, DC; and, as executive director of national cooperative development entities. She has managed the legislative agenda at the state and national levels for cooperative trade associations representing all sizes and sectors of the United States economy including Fortune 500 agriculture cooperatives and minority-owned catering businesses; farm credit banks and consumer credit unions; New York City and rural, senior housing; urban food stores and rural energy services.

Appendix H

Roundtable on Science and Technology for Sustainability

Established in 2002, the National Academies' Roundtable on Science and Technology for Sustainability provides a forum for sharing views, information, and analyses related to harnessing science and technology for sustainability. Members of the Roundtable include senior decision makers from government, industry, academia, and non-profit organizations who deal with issues of sustainable development, and who are in a position to mobilize new strategies for sustainability.

The goal of the Roundtable is to mobilize, encourage, and use scientific knowledge and technology to help achieve sustainability goals and to support the implementation of sustainability practices. Three overarching principles are used to guide the Roundtable's work in support of this goal. First, the Roundtable will focus on strategic needs and opportunities for science and technology to contribute to the transition toward sustainability. Second, the Roundtable will focus on issues for which progress requires cooperation among multiple sectors, including academia, government (at all levels), business, nongovernmental organizations, and international institutions. Third, the Roundtable will focus on activities where scientific knowledge and technology can help to advance practices that contribute directly to sustainability goals, in addition to identifying priorities for research and development (R&D) inspired by sustainability challenges.

In the summer of 2005, the Roundtable co-chairs convened a meeting with select leaders from the private sector, state government, nongovernmental organizations, academia, and the National Academies to help develop a strategic outlook for the second phase of the Roundtable. Meeting participants suggested a number of potential topics and modes of operations for the Roundtable. These ideas were used by the Roundtable co-chairs and staff to develop an action plan for activities to be undertaken by the Roundtable over the next few years.

To date, the Roundtable has explored topics such as linking knowledge with action for sustainable development, environmental regulation and its alternatives, sustainability indicators, rapid urbanization, and rebuilding the Gulf Coast Region in a sustainable manner. Major activities currently are being planned to examine the effectiveness of public-private partnerships for sustainability, certification of sustainable goods and services, urban environmental sustainability, sustainable energy, food security, and to discuss federal research and development activities to address selected high priority challenges to sustainability.

For Additional Information

For more information about the Roundtable, please contact Marina Moses, Director of the National Academies' Roundtable on Science and Technology for Sustainability, at *mmoses@nas.edu* or 202-334-2143.

Science and Technology for Sustainability Roundtable Membership

Emmy Simmons (Co-Chair) Assistant Administrator for Economic Growth, Agriculture, and Trade (retired), USAID

Thomas Graedel (Co-Chair) (NAE)
Clifton R. Musser Professor of Industrial Ecology, Yale University

Matt Arnold
Partner
PricewaterhouseCoopers

Ann M. Bartuska
Acting Deputy Under Secretary for Natural Resources and Environment, U.S. Department of Agriculture*

Arden Bement (NAE)
Director
National Science Foundation*

Michael Bertolucci
President
Interface Research Corporation

Nancy Cantor
President and Chancellor
Syracuse University

John Carberry
Director of Environmental Technology (retired)
DuPont

Leslie Carothers
President
Environmental Law Institute

William Clark (NAS)
Harvey Brooks Professor of International Science, Public Policy, and Human Development
Harvard University

Glen T. Daigger (NAE)
Senior Vice President and Chief Technology Officer
CH2M HILL

Patricia Dehmer
Acting Director
Office of Science
U.S. Department of Energy*

Sam Dryden
Managing Director
Wolfensohn & Company

Nina Fedoroff (NAS)
Science and Technology Advisor to the U.S. Secretary of State
U.S. State Department*

Marco Ferroni
Executive Director
Syngenta Foundation

Mohamed H. A. Hassan
Executive Director
The Academy of Sciences for the Developing World (TWAS)

Neil Hawkins
Vice President for Sustainability
The Dow Chemical Company

Geoffrey Heal
Garrett Professor of Public Policy and Business Responsibility
Graduate School of Business
Columbia University

Catherine (Katie) Hunt
Corporate Sustainability Director
Rohm and Haas

Lek Kadeli
Acting Assistant Administrator
Office of Research and Development
U.S. Environmental Protection Agency*

Jack Kaye
Associate Director
Research of the Earth Science Division
National Aeronautics and Space Administration*

Gerald Keusch (IOM)
Assistant Provost, Medical Campus
Associate Dean, School of Public Health
Boston University

Suzette Kimball
Acting Director
U.S. Geological Survey*

Kai Lee
Program Officer
Conservation and Science Program
Packard Foundation

Thomas E. Lovejoy
Biodiversity Chair
The H. John Heinz III Center for Science, Economics and the Environment

Pamela Matson (NAS)
Dean, School of Earth Sciences
Goldman Professor of Environmental Studies
Stanford University

J. Todd Mitchell
Chairman
Board of Directors
Houston Advanced Research Center

M. Granger Morgan (NAS)
Professor and Head
Department of Engineering and Public Policy
Carnegie Mellon University

Prabhu Pingali (NAS)
Head
Agricultural Policy and Statistics
Agriculture Development Division
Bill and Melinda Gates Foundation

Per Pinstrup-Andersen
H.E. Babcock Professor of Food, Nutrition and Public Policy, Nutritional Sciences
Professor, Applied Economics and Management
Cornell University

Christopher Portier
Associate Director
National Institute for Environmental Health Sciences (NIEHS)

Harold Schmitz
Chief Science Officer
Mars Inc.

Robert Stephens
International Chair
Multi-State Working Group on Environmental Performance

Denise Stephenson Hawk
Chairman
The Stephenson Group, LLC

Dennis Treacy
Vice President
Environmental and Corporate Affairs
Smithfield Foods

Vaughan Turekian
Chief International Officer
The American Association for the Advancement of Science*

*Denotes ex-officio member

Staff

Marina Moses, Director, Roundtable on Science and Technology for Sustainability

Pat Koshel, Senior Program Officer

Derek Vollmer, Associate Program Officer

Kathleen McAllister, Research Associate

Emi Kameyama, Program Assistant